Introduction to Genetics

Sandra Pennington
Department of Genetics
University of Washington
Seattle, Washington

Blackwell
Science

Editorial Offices:
Commerce Place, 350 Main Street, Malden, Massachusetts 02148, USA
Osney Mead, Oxford OX2 0EL, England
25 John Street, London WC1N 2BL, England
23 Ainslie Place, Edinburgh EH3 6AJ, Scotland
54 University Street, Carlton, Victoria 3053, Australia
Other Editorial Offices:
Blackwell Wissenschafts-Verlag GmbH, Kurfürstendamm 57, 10707 Berlin, Germany
Blackwell Science KK, MG Kodenmacho Building, 7–10 Kodenmacho Nihombashi, Chuo-ku, Tokyo 104, Japan

Distributors:
USA
Blackwell Science, Inc.
Commerce Place
350 Main Street
Malden, Massachusetts 02148
(Telephone orders: 800-215-1000 or 781-388-8250; fax orders: 781-388-8270)

Canada
Login Brothers Book Company
324 Saulteaux Crescent
Winnipeg, Manitoba, R3J 3T2
(Telephone orders: 204-224-4068)

Australia
Blackwell Science Pty, Ltd.
54 University Street
Carlton, Victoria 3053
(Telephone orders: 03-9347-0300; fax orders: 03-9349-3016)

Outside North America and Australia
Blackwell Science, Ltd.
c/o Marston Book Services, Ltd.
P.O. Box 269
Abingdon
Oxon OX14 4YN
England
(Telephone orders: 44-01235-465500; fax orders: 44-01235-465555)

Acquisitions: Nancy Hill-Whilton
Development: Jill Connor
Production: Irene Herlihy
Manufacturing: Lisa Flanagan
Interior design by Colour Mark
Cover design by Madison Design
Typeset by Best-set Typesetter Ltd., Hong Kong
Printed and bound by Capital City Press

Printed in the United States of America
00 01 02 03 5 4 3 2 1

The Blackwell Science logo is a trade mark of Blackwell Science Ltd., registered at the United Kingdom Trade Marks Registry

Library of Congress Cataloging-in-Publication Data

Pennington, Sandra.
Introduction to genetics / Sandra Pennington.
p. cm. — (11th hour)
ISBN 0-632-04438-1
1. Genetics. I. Title. II. Series: 11th hour (Malden, Mass.)
QH430.P46 2000
576.5—dc21

99-25401
CIP

CONTENTS

11th Hour Guide to Success vii
Preface viii

1. Mendelian Inheritance 1
   1. The Reasons for Mendel's Success 1
   2. Dominance and Recessiveness 3
   3. The Principle of Segregation 6
   4. The Testcross 8
   5. The Principle of Independent Assortment 11
   6. Testcrosses Revisited 13
   7. Mendel and Meiosis 15

2. Probability and Statistics 22
   1. Probability and the Product Rule 22
   2. Sum Rule 24
   3. Conditional Probability 27
   4. Binomial Probability 29
   5. Significance Testing with the Chi-Square Analysis 30

3. Extensions of Mendelian Inheritance 38
   1. Pedigrees 38
   2. Incomplete Dominance and Codominance 43
   3. Lethal Alleles 45
   4. Multiple Alleles 47
   5. Gene Interaction 49
   6. Sex Linkage 52

FIRST MIDTERM EXAM 61

4. Linkage 65
   1. Background for Linkage 65
   2. Linkage and Crossing Over 67
   3. Genetic Maps 70
   4. Gene Order and Distance 73
   5. Interference 77

5. DNA, RNA, and Proteins 84
   1. DNA Structure 84
   2. Organization of Eukaryotic Chromatin and Chromosomes 87
   3. Replication 90

4. RNA and Transcription 93
5. Protein and Translation 97
6. Control of Gene Expression 102
7. DNA Repair Systems 106

6. Mutation 113

1. Mutation 113
2. Mutagens 116
3. Features of Mutation 119
4. Classification of Mutations 121

7. Chromosomal Aberrations 127

1. Polyploidy 127
2. Aneuploidy 130
3. Deletions 133
4. Duplications 136
5. Inversions 138
6. Translocations 140

SECOND MIDTERM EXAM 147

8. The Gene as a Functional Unit 151

1. One Gene–One Enzyme 151
2. Relationship Between Gene Sequence and Protein Sequence 155
3. Complementation Analysis 156

9. Techniques of Molecular Genetics 165

1. Recombinant DNA 165
2. Genomic and cDNA Libraries 169
3. Restriction Mapping and RFLP Mapping 171
4. Sequencing 174
5. Polymerase Chain Reaction and Sequence Tagged Sites 176

10. Population Genetics 185

1. Genotypic and Allelic Frequencies 185
2. Hardy-Weinberg Law 188
3. Testing for Hardy-Weinberg Equilibrium 190
4. Mutation and Migration 193
5. Random Mating 195
6. Genetic Drift 197
7. Natural Selection 200

FINAL EXAM 209

11TH HOUR GUIDE TO SUCCESS

The 11th Hour Series is designed to be used when the textbook doesn't make sense, the course content is tough, or when you just want a better grade in the course. It can be used from the beginning to the end of the course for best results or when cramming for exams. Both professors teaching the course and students who have taken it have reviewed this material to make sure it does what *you* need it to do. The material flows so that the process keeps your mind actively learning. The idea is to cut through the fluff, get to what you need to know, and then help you understand it.

Essential Background. We tell you what information you already need to know to comprehend the topic. You can then review or apply the appropriate concepts to conquer the new material.

Key Points. We highlight the key points of each topic, phrasing them as questions to engage active learning. A brief explanation of the topic follows the points.

Topic Tests. We immediately follow each topic with a brief test so that the topic is reinforced. This helps you prepare for the real thing.

Answers. Answers come right after the tests; but, we take it a step farther (that reinforcement thing again), we explain the answers.

Clinical Correlation or Application. It helps immeasurably to understand academic topics when they are presented in a clinical situation or an everyday, real-world example. We provide one in every chapter.

Demonstration Problem. Some science topics involve a lot of problem solving. Where it's helpful, we demonstrate a typical problem with step-by-step explanation.

Chapter Test. For more reinforcement, there is a test at the end of every chapter that covers all of the topics. The questions are essay, multiple choice, short answer, and true/false to give you plenty of practice and a chance to reinforce the material the way you find easiest. Answers are provided after the test.

Check Your Performance. After the chapter test we provide a performance check to help you spot your weak areas. You will then know if there is something you should look at once more.

Sample Midterms and Final Exams. Practice makes perfect so we give you plenty of opportunity to practice acing those tests.

The Web. Whenever you see this symbol the author has put something on the Web page that relates to that content. It could be a caution or a hint, an illustration or simply more explanation. You can access the appropriate page through *http://www.blackwellscience.com*. Then click on the title of this book.

The whole flow of this review guide is designed to keep you actively engaged in understanding the material. You'll get what you need fast, and you will reinforce it painlessly. Unfortunately, we can't take the exams for you!

PREFACE

How many times have you been struggling with a new or difficult concept and just needed someone to explain it or phrase it a different way? This book will do that for genetics. By eliminating much of the peripheral details of a textbook it gives you a more concise resource. Whether you are trying to get through an undergraduate course or are studying for the GRE or MCAT, you can use this study guide to strengthen your understanding of basic genetics.

This study guide includes the central topics found in most introductory level genetics courses. You will find that the chapters are organized in such a way as to lead you from one topic to the next in a logical order, building on the knowledge you have gained and preparing you for the more complex ideas, such as linkage and population genetics. Some instructors might organize courses differently; however, with this study guide, that is not a problem. The Essential Background sections at the beginning of each chapter will tell you what concepts are important to the understanding of the new material, allowing you to move freely throughout the guide. You can also use the Essential Background as a checklist to make sure you have fully mastered the previous material.

As with all coursework, the best way to prepare for an exam is to practice, practice, practice; therefore, this study guide has numerous test questions after every major concept and at the end of each chapter. If this is not enough, there are mid-term and final exams that test your cumulative knowledge. The tests come in a variety of formats: true/false, multiple choices, short answers, and essay questions. Furthermore, answers are supplied for all questions, most with explanations.

This guide works best when used in conjunction with your textbook for supplemental figures. Additionally, there are complementary material and figures on the World Wide Web (http://www.blackwellscience.com). Just look for the at relevant locations.

This module would not exist without the assistance of my acquisitions editor Nancy Hill-Whilton and developmental editor Jill Connor. I am also grateful for the comments of Debbie Birnby, Ph.D., a great friend and insightful proofreader. I thank the following reviewers for their numerous helpful suggestions for improving the presentation and the coverage of this study guide. Special thanks go to the student reviewers. After all, this book is intended to help students: Arden Campbell, Iowa State University; Rick Duhrkopf, Baylor University; and Randy Krauss, Koren Mann, and Anthony Trombino, Boston University. Finally, words can not encompass the gratitude I have for all of Douglas' support.

Good luck with your studies.

Sandra Pennington, Ph.D.

Mendelian Inheritance

Inheritance of visible traits has been known since ancient times. For example, many domesticated species, like the dog, showed the signs of selective breeding by 8000 BC. Some of today's most important farming crops bear little resemblance to their wild forebears. Modern wheat is so different from its ancestor that it took detailed molecular analysis to discover the relationship. The sheep of 5000 BC were solely a source of food; their wool was too prickly to be used for clothing. In case this is not sufficient evidence to conclude that ancient peoples understood "like begets like," we have the images in the art of Ancient Egypt that show deliberate mating of select palm trees. Today, advances in medical genetics make news on a weekly basis. Geneticists are involved in many pursuits, ranging from identifying disease-causing genes to trying to cure those diseases through gene therapy. The basis for this field of study is the observation that most traits are heritable, including traits like eye color, which is biologically heritable, and the speaking of particular languages, which is culturally heritable. Genetics is the study of biologically heritable traits.

ESSENTIAL BACKGROUND

- **Organization of eukaryotic cells**
- **Basics of cellular reproduction (mitosis and meiosis)**

TOPIC 1: THE REASONS FOR MENDEL'S SUCCESS

KEY POINTS

✓ *Who performed the critical experiments determining the rules governing inheritance of traits?*

✓ *What key features of these experiments made them so definitive?*

✓ *What is a cross? F_1? F_2? a true breeder? homozygote?*

Modern genetics was born in the 1860s when a Moravian monk named Gregor Mendel performed the first carefully designed matings between dissimilar organisms and analyzed the offspring. Mendel wanted to be able to predict what the progeny of specific matings would look like. If there were more than one possibility for the progeny's appearance, he also wanted to know the statistical relationship between the progeny types.

Mendel was not the first person to attempt these kinds of experiments, but he was the first to succeed, largely because of the care he took to analyze one or a few traits in painstakingly controlled matings and keep fastidious records of all results. Mendel began his experiments with true-breeding **(homozygous)** lines of the garden pea (*Pisum sativum*). **True breeding** means that the characteristics of each line are constant from one generation to the next. For example, tall plants produce tall progeny, which in turn produce tall progeny. Normally peas reproduce through self-fertilization, or **selfing**; Mendel took great care to prevent this when it would have interfered with the mating, or **cross**, he conducted. For each cross, Mendel kept careful records of the appearance of the parent plants and both the appearance and the numbers of each type of progeny. Finally, Mendel collected sufficient progeny from each mating to make statistically sound conclusions. This amounted to hundreds or thousands of progeny per trait analyzed. His conclusions are known as the **principles of inheritance**. The principles allow one to predict the appearance and the relative proportions of offspring.

Recognizing that the results of previous experiments by others had been too complex to interpret, Mendel began his experiments by crossing peas that differed in only a few traits at most. These dissimilar peas make up the **P (parental) generation**. Mendel recorded their progeny, called the **F_1 (first filial) generation**, and allowed them to make another generation by selfing. Their offspring, the **F_2 (second filial) generation**, exhibited remarkable characteristics, which Mendel was able to quantify, and which allowed him to devise rules that govern the inheritance of traits.

Topic Test 1: The Reasons for Mendel's Success

True/False

1. The matings Mendel conducted were between groups of vastly different pea plants, which he allowed to mate indiscriminately.
2. Mendel wanted to generate precise rules governing the transmission of visible traits.
3. Ten progeny would have been enough for Mendel's conclusions to be significant.

Multiple Choice

4. Genetics is the study of the inheritance of
   a. interesting traits.
   b. all traits.
   c. diseases.
   d. social customs.
   e. biological traits.
5. A true-breeding pea plant
   a. can be pollinated only at a particular time of day.
   b. looks like its parents and gives offspring that look exactly like it.
   c. can only be mated to its own kind.
   d. only breeds true in alternate generations.
   e. does none of the above.

Short Answer

6. Describe any three of the important features of Mendel's experiments and explain how they facilitated his analysis.

Topic Test 1: Answers

1. **False.** The matings were between individual plants that differed in appearance in only one or a few traits. The matings were only allowed to occur between the intended plants.
2. **True.** He wanted to be able to predict accurately what progeny would result from any mating.
3. **False.** Ten is well within the range of numbers where random fluctuation could have a profound effect on the observed ratio. We will discuss this concept in greater detail in Chapter 2.
4. **e.** Traits that have a biological, rather than social, basis are the concern of geneticists.
5. **b.** True-breeding, also called homozygous, organisms have exactly this property: the appearance of the particular trait is the same in all generations, so long as all matings occur within that line of peas.
6. Few traits: By limiting the experiments to analyzing a few traits at a time, Mendel greatly decreased the complexity of the results by decreasing the number of different types of progeny produced.

   Controlled matings: By preventing self-fertilization and accidental cross-pollination, Mendel reduced the complexity of the progeny, generating fewer types from known parents.

   Good records: Mendel kept track of which plants were crossed to generate particular progeny and carefully counted and categorized all progeny.

   True breeders: The exclusive use of pure lines ensured the predictable outcomes by preventing masked traits in the parents. (This point will become more clear in Topic 2.)

   Large numbers of progeny per experiment: This gave Mendel the strength of numbers that enabled him to make a statistically significant conclusion.

TOPIC 2: DOMINANCE AND RECESSIVENESS

KEY POINTS

- ✓ *What are dominance and recessiveness?*
- ✓ *How are dominance and recessiveness determined?*
- ✓ *What are heterozygotes? hybrids? alleles? phenotypes? genotypes?*

The concepts of dominance and recessiveness are derived from the results of many matings between pea plants that differ in one or a few characters. For example, pea seeds can be round or wrinkled. Round and wrinkled are **phenotypes** of pea seeds, meaning they are what pea seeds look like. Mendel's crosses involving these two traits are diagrammed in **Figure 1.1**. The parent (P) plants were true-breeding (homozygous) round-seed pea plants and true-breeding wrinkled-seed pea plants. Remarkably, *all* of their F_1 were round seeds. (The seeds borne on the maternal plant are the offspring of the mating and exhibit the appropriate F_1 phenotype.) None had blended phenotypes; none were half round and half wrinkled; all were entirely round. This is the primary evidence for the concepts of **dominance** and **recessiveness**. Round is dominant because these F_1 plants are round, despite having both round information from the

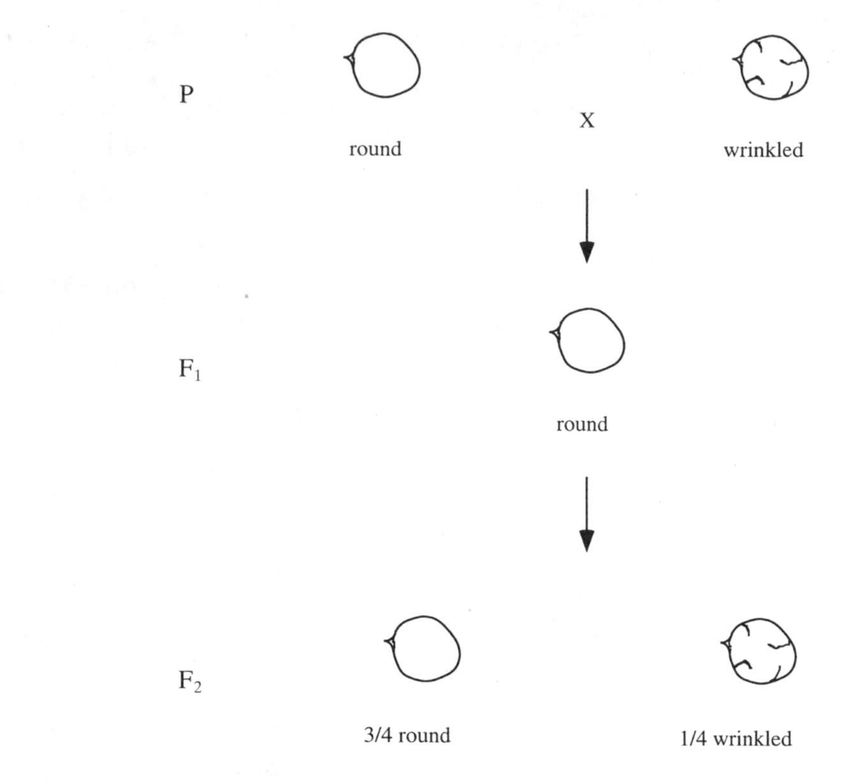

Figure 1.1 Monohybrid cross.

round parent and wrinkled information from the wrinkled parent (i.e., they are **heterozygous**). The proof that the F_1 plants contain wrinkled information can be observed in the progeny created by selfing the F_1 generation: In the F_2 seeds, the wrinkled phenotype reappears (described in Topic 3). Since all of the F_1 were round seeds, the round trait must be dominant to the wrinkled-seed trait, which is recessive.

Mendel called the information for each trait **unit factors of inheritance**; they are now known as **alleles**. Round and wrinkled are alleles (different versions) of the same gene, the seed-shape gene, and we can give these alleles symbols that will remind us of their relationship. One convention for the symbols is to abbreviate the name of the recessive trait, often using just the initial letter. We could call the wrinkled allele *w*, using the lowercase letter to indicate its recessiveness. Round is a different allele of the same gene (seed shape) so it also is abbreviated with *W*, but notice the capitalization of this letter to indicate round's dominance. Each of the true-breeding parents has two of the same kind of alleles. We say the parents are homozygous and we could abbreviate this *WW* for round and *ww* for wrinkled. *WW* and *ww* are **genotypes**, the specific alleles possessed by an organism. Each parent passes one allele to each of their offspring at fertilization. Thus, the F_1 seeds have the genotype *Ww*. Since these are dissimilar alleles, the F_1 are said to be heterozygous or **hybrid** (**monohybrid** in this case because they are hybrid for a single gene). An important point to remember is that dominance and recessiveness can only be determined in a heterozygote. If a heterozygote cannot be positively identified, for example, as the progeny of a cross between two dissimilar homozygotes, then the dominance-recessiveness relationship between the alleles cannot be determined.

Topic Test 2: Dominance and Recessiveness

True/False

1. The genotype is the organism's appearance.

2. The dominant trait is the one the heterozygote exhibits.

3. Alleles are the different forms of genes.

Multiple Choice

4. If a cross between a true-breeding green-seed pea and a true-breeding yellow-seed pea produced yellow progeny, then
   a. yellow is dominant.
   b. green is dominant.
   c. it is impossible to tell which is the dominant allele.
   d. the next time these parents are crossed, green-seed progeny will result.

5. If a cross between a tall pea plant and a short pea plant produced 490 tall and 478 short offspring, what is the dominant trait?
   a. Neither is dominant.
   b. Tall is dominant.
   c. Short is dominant.
   d. It is impossible to tell which is the dominant allele.

Short Answer

6. Briefly define homozygous, true breeder, F_1, allele, phenotype, genotype, heterozygous, hybrid, dominant, and recessive.

Topic Test 2: Answers

1. **False.** The phenotype is the organism's appearance. The genotype is the organism's allele constitution.

2. **True.** Heterozygotes are useful for determining dominance, and the trait the heterozygotes show is the dominant one (specific exceptions will be described in Chapter 3).

3. **True.** Alleles are to genes what colors are to automobiles. Both the round and wrinkled alleles are versions of the seed-shape gene, much as a blue and a red truck are different forms of trucks.

4. **a.** The yellow progeny must be heterozygous, since its parents were both true breeders and were dissimilar. The phenotype of the heterozygotes indicates dominance.

5. **d.** It is impossible to tell, because it is impossible to identify the heterozygote in this cross. At least one of the two peas being mated appears not to be a true breeder, since the progeny exist in two phenotypic classes (also see Topic 4).

6. Homozygous means having two of the same allele for a particular gene. True breeder is synonymous with homozygote and refers to the constancy of traits over the course of generations. F_1 is the generation and the members of it that are the offspring of two parental lines. An allele is one of the variant forms of a gene. Phenotype is the appearance of an organism. Genotype is that organism's allele constitution. Heterozygous is having two different alleles of a particular gene. Hybrid is a heterozygote or the progeny from the cross of two dissimilar parental lines. Dominant is an allele or a phenotype that

masks another allele or phenotype (of the same gene) in the heterozygote. Recessive is the allele and trait that is not seen (i.e., is masked) in the heterozygote.

TOPIC 3: THE PRINCIPLE OF SEGREGATION

KEY POINTS

✓ *What does segregation mean?*

✓ *What does a 3:1 phenotypic ratio mean?*

✓ *What does a 1:2:1 ratio mean?*

In the F_2 generation of the cross described in Topic 2, Mendel counted 5,474 round and 1,850 wrinkled seeds. There are four important points to understand from these progeny. The first is the reappearance of the wrinkled trait. In order to reappear in the F_2 generation, the wrinkled trait must have been present but completely masked in the F_1 generation. Notice also that the wrinkled trait appears in this generation because it is part of the progeny from a selfed heterozygote. The cross of two heterozygotes will produce progeny with the recessive phenotype, at a particular frequency. This frequency is the second point: The ratio between progeny classes is 3:1. When any two contrasting, true-breeding parents were crossed, and Mendel tested seven characters, the ratio of dominant to recessive in the F_2 generation was always 3:1. A broader implication of this finding is that 3:1 is the **phenotypic ratio** that should be expected when crossing all heterozygotes, like the F_1. (At a later time, this statement will be softened, but for now it is true.) The 3:1 ratio arises because alleles are passed on to the progeny at equal frequency. The **principle of segregation** states that alleles will separate from each other during the production of gametes (the cells used for sexual reproduction in eukaryotic organisms, e.g., sperm and oocytes) so that they are equally transmitted to the progeny (see also Punnett squares in Topic 4).

The third point concerns the number of individuals in each progeny class: thousands! This was extremely important to Mendel's success because the larger the number of individuals in each phenotypic class, the less random variation will affect the ratio between the classes. Put another way, chance will influence the results of any experiment. But the greater the number of items being measured, the less of an effect chance will have on the relationship between those items. Note that 3:1 is only a ratio, based on the actual data Mendel collected. He counted hundreds or thousands of F_2 for each of the seven characteristics. This large amount of data added statistical weight to his analysis. If he had really counted only four peas, he would not have had much of an argument.

Finally, the fourth point is the identities of the F_2 offspring. Mendel wanted to know whether all of the F_2 seeds were like the parents (i.e., homozygotes) or whether some were like the F_1 (i.e., heterozygotes). To make this distinction, he allowed the F_2 to self-fertilize and analyzed the resulting F_3. The wrinkled F_2 made wrinkled F_3, so these must be homozygous like the original wrinkled parents. The round F_2 fell into two classes. One-third of the round F_2 had all round offspring, just like the original round-seed parent, so these F_2 offspring must be homozygous. The other two-thirds, however, were like the F_1, giving three-fourths round and one-fourth wrinkled offspring. Collectively that makes one-fourth homozygous round (*WW*), two-fourths heterozygous round (*Ww*), and one-fourth homozygous wrinkled (*ww*). This ratio, 1:2:1, is the **genotypic ratio** for the F_2 generation in a simple (single-gene) Mendelian cross. The two genotypic classes that display the dominant phenotype comprise (1/4 + 2/4) 3/4 of the offspring when grouped by phenotype. Hence, the monohybrid F_2 phenotypic ratio is 3:1, where 3/4 of the offspring have the dominant phenotype (genotypically *W*-, in which the dash means *W* or *w*) and 1/4 have the recessive phenotype (genotypically *ww*).

Topic Test 3: The Principle of Segregation

True/False

1. For a simple Mendelian trait, the progeny from a cross of two heterozygotes will fall into two phenotypic classes of roughly equal number.
2. A single phenotype can be caused by more than one genotype.
3. The progeny of a cross between two dissimilar true breeders are also true breeders.

Multiple Choice

4. For a simple Mendelian trait, a selfed heterozygote can be expected to yield progeny phenotypes in the ratio
   a. 1:1.
   b. 1:1:1:1.
   c. 1:2:1.
   d. 3:1.
   e. 9:3:3:1.
5. In a single-gene Mendelian cross, what fraction of the F_2 generation are true breeders?
   a. None
   b. One-fourth
   c. One-half
   d. Three-fourths
   e. All
6. Mating two individuals results in a 3:1 phenotypic ratio in the progeny. The genotypes of the mating are
   a. *Aa* × *AA*.
   b. *AA* × *aa*.
   c. *Aa* × *aa*.
   d. *Aa* × *Aa*.

Short Answer

7. A cross between a true-breeding green-seed pea and a true-breeding yellow-seed pea produced yellow progeny.
   a. What trait is dominant?
   b. If the F_1 are selfed, what proportion of the F_2 generation will be green?

Topic Test 3: Answers

1. **False.** The cross of two heterozygotes will yield two phenotypic classes of progeny; three-fourths exhibit the dominant phenotype and one-fourth exhibit the recessive phenotype.

The 1:1 ratio described in this question is characteristic of a different type of cross described in Topic 4.

2. **True.** The dominant phenotype is caused by two genotypes: homozygous dominant and heterozygous.
3. **False.** The progeny of dissimilar true breeders will be heterozygous and therefore will give progeny exhibiting both phenotypes. This is not "true breeding."
4. **d.** The typical ratio for the progeny of a selfed heterozygote is 3:1, where the larger class shows the dominant phenotype. The other ratios listed are ratios expected from other matings that will be discussed later.
5. **c.** The homozygous round progeny are one-fourth of the total and the homozygous wrinkled progeny are one-fourth of the total. Therefore, one-half of the progeny are true breeders (homozygous).
6. **d.** A 3:1 phenotypic ratio indicates heterozygosity of both parents. The first two matings produce progeny of a single phenotype. The third is a testcross, a type of mating that will be covered in the next topic.
7. Yellow is dominant because that is what the heterozygous F_1 look like. One-fourth of the F_2 generation will be green, because green is recessive, so only the offspring that are homozygous recessive will be green.

TOPIC 4: THE TESTCROSS

KEY POINTS

✓ *How can the genotypes of individuals be determined?*

✓ *What exactly is a testcross?*

✓ *What does a 1:1 ratio signify?*

✓ *What is a Punnett square?*

When a single gene is considered, any organism expressing the recessive phenotype is homozygous for the recessive allele. For example, the wrinkled peas are *ww*. Yet the genotype of any randomly chosen dominant organism is unknown. It could be either homozygous (*WW*) or heterozygous (*Ww*). The surest way to distinguish between these two possibilities is to perform a **testcross** of the unknown organism. To perform a testcross, the organism whose genotype is unknown is crossed to a recessive homozygote. There are two possible outcomes to the cross, depending on whether the unknown organism is homozygous or heterozygous. If all the progeny have the dominant phenotype, then they are heterozygotes and their unknown parent was homozygous for the dominant trait. Alternatively, if the progeny consist of roughly equal numbers of dominant and recessive individuals, then the unknown parent was heterozygous. The testcross is used extensively by geneticists for precisely this purpose. The technique also points out how easy it can be to reason through genotype assignments.

A testcross of a heterozygous F_1 round pea to a wrinkled (homozygous recessive) one produces round and wrinkled progeny *in equal proportions*. That is, the ratio is 1:1. This proves two points. First, we are seeing once again that the F_1 plant is heterozygous: It has one allele specifying the round trait and one allele specifying the wrinkled trait. The second point is that these alleles

	1/2 *W*	1/2 *w*
1/2 *W*	1/4 *WW* round	1/4 *Ww* round
1/2 *w*	1/4 *Ww* round	1/4 *ww* wrinkled

Figure 1.2 Punnett square of the monohybrid cross.

must be passed on to the progeny equally frequently. Understanding this point caused Mendel to propose the principle of segregation, which states that alleles will separate from each other during the production of gametes so that they are equally transmitted to the progeny. This was a remarkable statement for Mendel to make since he described meiosis four decades before it would be observed in a microscope.

Before we move on to another of Mendel's principles, there is a method of bookkeeping that is best introduced now. The technique is called the **Punnett square**, named after the geneticist who devised it, R. C. Punnett. Executed perfectly, it completely predicts all of the progeny expected for a particular cross. As an example, let us consider the selfing of the round F_1 generation (**Figure 1.2**). First we need to determine what kinds of gametes the F_1 will make. Recall that the F_1 generation's genotype is *Ww*. Mendel determined that alleles segregate away from each other during meiosis, so the *W* will segregate from *w*, and the F_1 plants will make two kinds of gametes: those that contain *W* and those that contain *w*. Since this is a self-fertilization, both parents will make these two kinds of gametes. The Punnett square is drawn by placing the gametes along both the top and left side of the square, one gamete in each column and row, as depicted in Figure 1.2. To determine the genotypes for all of the progeny of the selfing, copy the gametes each parent contributes to the progeny into each square. Some of the squares will show the same genotype; count these to determine the proportion of progeny possessing that genotype. For the example in Figure 1.2, the total genotypic ratio is 1 *WW* : 2 *Ww* : 1 *ww*. Phenotypes can be determined for each genotype as well, and subsequently counted for the phenotypic ratio (3 : 1 in Figure 1.2). The problem many students encounter executing a Punnett square is determining the correct gametes, so pay close attention to Mendel's discoveries and Topic 7, "Mendel and Meiosis," for guidelines in this matter.

Topic Test 4: Testcross

True/False

1. A testcross is the mating of a homozygous recessive organism to an organism showing the dominant phenotype in order to ascertain the latter's genotype.
2. Punnett squares do not account for all possible progeny.

Multiple Choice

3. If the progeny of a testcross are divided equally into two phenotypic classes, what are the parental genotypes?

a. $BB \times bb$
b. $Bb \times bb$
c. $Bb \times BB$
d. $Bb \times Bb$
e. Both a and b
f. All of the above are possible.

4. If the progeny of a cross exhibit the same phenotype, what are the parental genotypes? (This might not be a testcross.)
a. $DD \times dd$
b. $Dd \times dd$
c. $Dd \times DD$
d. $Dd \times Dd$
e. Both a and c
f. All of the above

Short Answer

5. Draw a Punnett square for each of the following crosses. Determine genotypic and phenotypic ratios for each.
a. $WW \times ww$
b. $Ww \times Ww$
c. $Ww \times ww$

Topic Test 4: Answers

1. **True.** One parent must be homozygous recessive and the other must show the dominant phenotype to be a testcross.

2. **False.** When executed correctly, the Punnett square is a precise method for determining all progeny. Most errors are one of two types: procedural or conceptual. The conceptual errors consist of using the wrong gametes for the cross. The procedural errors are carelessness in filling in the boxes.

3. **b.** This question asks you to recognize a 1 : 1 ratio as being the result of testcrossing a heterozygote.

4. **e.** This question asks you to recognize that the results of some crosses are ambiguous. Both of these matings satisfy the requirement of producing only dominant offspring the same way: One parent is homozygous dominant.

5. In cross a, one parent produces only *W* gametes and the other parent produces only *w* gametes. All of their offspring are *Ww*. In cross b, both parents produce two kinds of gametes, *W* and *w*, in equal proportions. Their offspring are one-fourth *WW*, one-half *Ww*, and one-fourth *ww* for a phenotypic ratio 3 *W* type : 1 *w* type. In cross c, the heterozygous parent makes *W* and *w* gametes and the homozygous recessive parent makes *w* gametes, only. Their progeny are one-half *Ww* and one-half *ww* progeny (half of each phenotype).

TOPIC 5: THE PRINCIPLE OF INDEPENDENT ASSORTMENT

KEY POINTS

✓ *What is independent assortment?*

✓ *What is the evidence for independent assortment?*

✓ *What does a 9:3:3:1 ratio signify?*

Mendel wondered what would happen if more than one character varied in the crosses. He assayed several different combinations of characters; we will look closely at one. In this experiment, he started with homozygous lines of yellow round peas and green wrinkled peas. When he crossed these two lines, they created F_1 that were yellow round seeds. This shows that yellow and round are dominant to green and wrinkled, respectively. The F_1 were selfed and the resulting F_2 generation fell into four classes, all four possible combinations of the original phenotypes. The actual numbers and types of progeny are as follows:

315 yellow round

101 yellow wrinkled

108 green round

32 green wrinkled

The surprise in these progeny is that there are two new types of plants with seeds not seen before in these lines: yellow wrinkled seeds and green round seeds. The ratio of these progeny is 9.8:3.2:3.4:1, or approximately 9:3:3:1. This is the expected ratio based on the known dominance and an assumption that the characters (color and texture) assort independently. **Independent assortment** means the color character is passed on to offspring independent of the texture character. To see this, consider the Punnett square for the selfing of the F_1 generation. First, let yellow be *G* and green, *g*; round is *W* and wrinkled, *w*. The P generation is *GGWW* crossed with *ggww*. All of their offspring are *GgWw* because each parent makes only one type of gamete. Each gamete must contain one allele of the shape gene and one allele of the color gene. Both parents are homozygous so all gametes from each parent will be the same. The *GGWW* parent only makes *GW* gametes and the *ggww* parent only makes *gw* gametes. Fertilization between *GW* and *gw* results in *GgWw*, a double heterozygote, or **dihybrid**. This cross of F_1 with F_1 is sometimes referred to as a **dihybrid cross**. To make the Punnett square representing the selfed F_1, we need to determine the F_1 dihybrid's gametes. At this point, Mendel reasoned that not only were the alleles segregating apart (see Topic 3) but also the genes were assorting independently.

If he is correct, then the F_1 dihybrid must make four kinds of gametes. The *G* allele must segregate from the *g* allele and, independently, the *W* allele must segregate from the *w* allele. The four possible combinations of these four alleles are *GW*, *Gw*, *gW*, and *gw*. These form the top and left sides of the Punnett square (**Figure 1.3**). (The fact that there are 16 progeny boxes does *not* mean that there are only 16 offspring nor that there are 16 phenotypes or genotypes.) Notice that some of the genotypes appear more than once in the square. Count all of the instances of a particular genotype (for example, *GgWw* appears four times), and the resulting number divided by 16 is the expected frequency of that *genotype* in the F_2 generation (e.g., 4/16), provided the genes assort independently. The famous ratio 9:3:3:1 is the *phenotypic* ratio for these offspring. Count for yourself the nine boxes that represent yellow round progeny.

	1/4 *GW*	1/4 *Gw*	1/4 *gW*	1/4 *gw*
GW 1/4	*GGWW* yellow round 1/16	*GGWw* yellow round 1/16	*GgWW* yellow round 1/16	*GgWw* yellow round 1/16
Gw 1/4	*GGWw* yellow round 1/16	*GGww* yellow wrinkled 1/16	*GgWw* yellow round 1/16	*Ggww* yellow wrinkled 1/16
gW 1/4	*GgWW* yellow round 1/16	*GgWw* yellow round 1/16	*ggWW* green round 1/16	*ggWw* green round 1/16
gw 1/4	*GgWw* yellow round 1/16	*Ggww* yellow wrinkled 1/16	*ggWw* green round 1/16	*ggww* green wrinkled 1/16

Figure 1.3 Punnett square of the dihybrid cross.

The most striking aspect to these progeny is the existence of yellow wrinkled and green round F_2. These progeny are unlike any plants seen before in this experiment in that color and shape are combined differently in the F_2 generation than they were in the P generation. A less obvious aspect is that the frequency at which these exceptional progeny appear is consistent with *random* reassortment of the characters or genes. If the gametes that form these progeny (*gW* and *Gw*) did not occur at the same frequency as the other two gametes (i.e., the "parental" gametes, *GW* and *gw*), a ratio other than 9:3:3:1 would have occurred. (In Chapter 4, we will examine the situation where reassortment is not random. Suffice it to say now that we can recognize non-random assortment of genes by an altered 9:3:3:1 dihybrid ratio.) Mendel recognized the randomness that is needed to create this distinctive ratio and correctly inferred from it the phenomenon that would be described by his **principle of independent assortment**. This principle states that the fate of one character (seed color) is independent of the other character (seed shape). In modern terms, we say that the segregation of alleles of one gene is independent of that of alleles of a second gene. The proof of this statement is the appearance of non-parental combinations of phenotypes in the proportions 3/16 green round seeds and 3/16 yellow wrinkled seeds in the F_2 generation.

Topic Test 5: Independent Assortment

True/False

1. The green round progeny in the cross discussed in this topic are clearly produced by independent assortment of genes.
2. The principle of independent assortment means that green and yellow will sometimes appear in the same seed.

Multiple Choice

3. The dihybrid ratio 9:3:3:1 refers to
   a. genotypes.
   b. phenotypes.

c. gametes produced by the dihybrid.
d. the number of generations needed to get a heterozygote.

4. The cross (by genotype) that produces progeny in the ratio 9:3:3:1 can be generalized:
   a. *AaBb* × *AaBb*
   b. *AAbb* × *aaBB*
   c. *AaBb* × *aabb*
   d. *AAbb* × *AaBb*
   e. *AAbb* × *aabb*

Short Answer

5. Describe how the results of the dihybrid cross proved the principle of independent assortment.

Topic Test 5: Answers

1. **True.** The fact that the green round progeny are not exactly like either one of the original true-breeding parents, and their frequency, indicates independent assortment. If there were no independent assortment, there would not be F_2 of this type. Independent assortment is responsible for the green round and the yellow wrinkled F_2.
2. **False.** Green and yellow will not appear in the same seed because they are alleles of the same gene and yellow is dominant. Independent assortment is a concept that applies to two or more genes, not to alleles of a single gene.
3. **b.** This famous ratio refers to the phenotypic classes produced by selfing a dihybrid. The dihybrid makes four kinds of gametes and there are nine genotypes represented by the progeny.
4. **a.** This famous ratio refers to the phenotypic classes produced by selfing a dihybrid. Cross b produces a dihybrid like the one that can produce the ratio in question. Cross c is the testcross of a dihybrid, and produces four classes of progeny in the ratio 1:1:1:1. Cross d produces two classes of progeny (in the ratio 3:1 with B varying). Cross e is the testcross of a monohybrid (b doesn't vary).
5. Parental traits were partitioned to produce two novel classes of progeny and they appeared at the frequency expected for random reassortment.

TOPIC 6: TESTCROSSES REVISITED

KEY POINTS

✓ *What happens when a dihybrid is testcrossed?*

✓ *What does the ratio 1:1:1:1 mean?*

The testcross is useful for unmasking recessive alleles in a plethora of situations, including organisms that may be heterozygous for two genes. Consider the yellow round progeny in the previous topic. Each of these peas could have any of four possible genotypes: *GGWW*, *GgWW*, *GGWw*, or

Table 1.1 Test Crosses of Two Yellow Round Peas

Yellow round		Homozygous recessive (green wrinkled)	Progeny
GGWW	×	*ggww*	All yellow round
GgWw	×	*ggww*	1/4 yellow round 1/4 yellow wrinkled 1/4 green round 1/4 green wrinkled

GgWw. There is no way to tell the genotype of a yellow round offspring simply by looking at it. Crossing the yellow round pea plants among themselves will not help: Each yellow round mating could be any of 10 possible genotype combinations. And some of the combinations give the same results. Hence the testcross specifies use of a homozygous recessive individual as the known parent. This creature will only contribute recessive alleles to the progeny, so any recessive alleles present but masked in the unknown parent will be revealed by progeny with the recessive phenotype. For example, see **Table 1.1**. Note that the progeny of this second cross are in the ratio 1:1:1:1. This ratio is a clear indication that the cross was a testcross to a dihybrid with independent assortment occurring. A testcross differentiates among *GGWW*, *GGWw*, *GgWW*, and *GgWw*, and allows us to determine the genotype of an unknown parent. This holds true even with more than two genes segregating.

Topic Test 6: Testcrosses Revisited

True/False

1. The ratio 1:1:1:1 is the phenotypic result of testcrossing a dihybrid exhibiting independent assortment.
2. In most situations, it is better not to use a testcross to determine genotypes.

Multiple Choice

3. If Mendel wanted to determine the genotype of a yellow round pea, what would be the best plant to cross to it?
   a. *GgWw*
   b. *GGWW*
   c. *ggWW*
   d. *GGww*
   e. *ggww*
4. What phenotypic ratio of offspring should be expected from the cross *AaBb* × *AAbb*?
   a. 1:1:1:1
   b. 2:1
   c. 9:3:3:1
   d. 1:1
   e. 3:1

Short Answer

5. Predict the progeny obtained from testcrossing a green round pea plant. (How many possible genotypes do you need to consider?)

Topic Test 6: Answers

1. **True.** Only a dihybrid will give four classes of progeny in a testcross. The equal proportions result from independent assortment.
2. **False.** The testcross is the method of choice for this purpose.
3. **e.** If the genotype of an individual is to be tested, the best cross to perform is the testcross, to a homozygous recessive. All of the other crosses will allow potential recessive alleles in the yellow round plant to remain masked.
4. **d.** Consider each gene separately. *Aa* × *AA* will only produce progeny with the A phenotype. *Bb* × *bb* is the testcross of a monohybrid, so will produce a 1 : 1 ratio. One class will have genotypes *AABb* and *AaBb* and the other class will be composed of *AAbb* and *Aabb* individuals.
5. Green round could be *ggWW* or *ggWw*; therefore, there are two possible outcomes of a testcross. If the unknown parent is *ggWW*, all the offspring will be green round seeds. If the unknown parent is *ggWw*, half of the offspring will be green round and half will be green wrinkled seeds.

TOPIC 7: MENDEL AND MEIOSIS

KEY POINTS

✓ *How do sexually reproducing organisms make gametes?*

✓ *In what genetic ways are gametes different from the other cells in the organism?*

✓ *What (microscopically) visible parts of a cell are responsible for the inheritance of Mendel's traits?*

A **gene** is a sequence of DNA that usually encodes a protein and occupies a specific site in a chromosome, and can also be called a **locus**. The seed-shape trait is one example of a gene. In most organisms, each gene exists in two copies, which are called **alleles**. Alleles can be dominant or recessive but can also be described as **wild type**, abbreviated **wt** or given the allele symbol "+." This name is usually given to the most common allele (the most common version of the trait), and is often the dominant allele, although it could be the recessive allele if that allele is more common. The pairs of alleles exist on pairs of chromosomes called **homologs**; there is one allele per homolog in the same position on both. In most organisms, including people, there are two homologs of every chromosome. They resemble tiny rods and are visible briefly during the cell cycle during mitosis in most **(somatic)** cells and during meiosis in sex cells **(germ cells)**. **Mitosis** is the nuclear division that makes cells identical: daughter cells identical to each other and to their mother cell. It occurs after the cell has copied all of its DNA during S phase; the copies are called **sister chromatids**. During M phase, the cell divides the identical sister chromatids apart to make two identical daughter cells. The homologs do not separate from each other, and this is what makes the daughter cells identical. Contrast mitosis with **meiosis**,

where the goal is to reduce the cell's DNA content by half, without losing any genes. It is a precise division in which the daughter cells keep only one member of each allele pair. This is done by segregating homologs apart in one nuclear division, then separating the sister chromatids apart in a second division. Thus, diploid organisms (organisms that have two homologs per chromosome) can make haploid sex cells that will fuse at fertilization to regenerate the diploid. The first observation of this behavior of chromosomes in meiosis was made in 1902 when Walter Sutton and Theodor Boveri, working independently, observed chromosomes at meiosis and realized that they behaved just like Mendel's "particles in gametes."

Early in meiosis, prophase I, the homologs come together and recombine, meaning the homologs exchange pieces of themselves with each other. This is necessary for the successful completion of meiosis and also facilitates independent assortment, as will be described in the chapter on linkage (Chapter 4). (It is worth noting here that recombination is extremely more rare in mitosis than in meiosis.) In metaphase I the paired homologs have lined up in the middle of the cell in preparation for separating away from each other in anaphase I. The subsequent telophase is very short in most organisms and is usually not followed by an interphase. An often brief prophase II is followed by metaphase II when the homologs, now technically in haploid cells, line up once again, this time to separate their sister chromatids in anaphase II. The cytoplasm divides along with the nuclei, and at the end of meiosis there are four haploid cells where once there was one diploid cell. The haploid cells are not identical to each other, nor to the mother cell; recall that the mother cell has two times as much DNA as the gametes. This is the strength of meiosis: making as many different combinations of alleles of genes as possible. Specifically, what Sutton and Boveri observed was that (1) chromosomes exist in pairs, just like Mendel's genes exist in pairs of alleles; (2) homologs segregate into gametes, just like Mendel's alleles of genes; and (3) different chromosome pairs assort independently from one another, just like Mendel's genes. They concluded that genes must be parts of chromosomes. Nearly 10 years later this **chromosomal theory of inheritance** was proved by the famous geneticist Thomas H. Morgan. (Meiosis will be revisited in Chapter 4.)

Topic Test 7: Mendel and Meiosis

True/False

1. Chromosomes and characters behave similarly with regard to inheritance at meiosis.
2. Mitosis makes genetically identical daughter cells.

Multiple Choice

3. Homologs are separated from each other
   a. during meiosis I.
   b. during meiosis II.
   c. during mitosis.
   d. during meiosis II and mitosis.
   e. never.
4. The stage of meiosis when pairing and recombination occurs is
   a. prophase I.
   b. metaphase I.

c. anaphase I.
d. prophase II.
e. metaphase II.

Short Answer

5. Draw a picture of the chromosomes of a *Ww* pea. Show a few schematics of these chromosomes going through meiosis.

Topic Test 7: Answers

1. **True.** The homologs of each chromosome appear to separate independently of the homologs of other chromosomes, just as the alleles of each gene separate independently of other genes' alleles.
2. **True.** The daughter cells produced by mitosis are genetically identical because the homologs are not segregated; the sister chromatids are.
3. **a.** This is the separation that makes meiosis unique and the gametes haploid.
4. **a.** Prophase I is temporally the longest phase of meiosis because of the large number of events that have to be completed before the metaphase I division can occur.
5. 

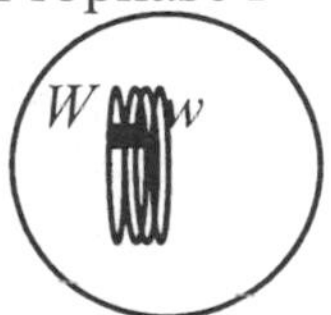

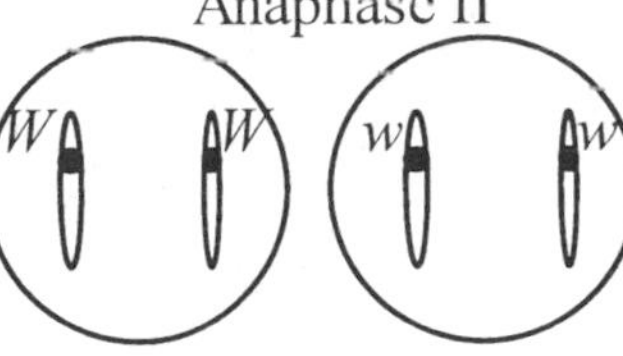

IN THE CLINIC

Large numbers of human traits obey Mendel's principles. Some well-known traits are caused by recessive or dominant alleles. Cystic fibrosis is a relatively common disease of the respiratory and digestive systems that causes premature death in recessive homozygotes. Among the white population living in the United States, approximately 1 of every 25 people is heterozygous for the disease allele. The inability to taste the chemical phenylthiocarbamide (PTC) is caused by homozygosity for a recessive allele. Chin clefts are caused by a dominant allele. A dominant allele of another gene causes the chin to tremble in response to anxiety or distress. General Tom Thumb (Charles S. Stratton) was a famous performer with P. T. Barnum's circus. His name and act were based on his short stature; he measured 3 feet 2 inches tall, presumably due to homozygosity for a recessive allele that prevented production of sufficient amounts of pituitary growth hormone. Stratton's parents were first cousins, a mating that increases the likelihood of producing homozygous recessive children (see Chapter 3). This points to one of the common modern uses of genetic information, genetic counseling, where knowledge of the genetic basis for diseases can help members of affected families know the likelihood that they or their children will inherit the disease.

Table 1.2 Results for Five Matings of Specific Plants					
		PROGENY PHENOTYPES			
MATING	PARENTAL PHENOTYPES	WHITE SQUARE	RED SQUARE	WHITE ROUND	RED ROUND
1	red square × white square	0	761	0	254
2	red round × white square	210	202	196	199
3	red square × red round	49	158	54	153
4	red square × white square	225	231	76	80
5	red square × white round	206	200	0	0

DEMONSTRATION PROBLEM

Question: Flowers of a particular plant can be either red or white, and be square or round. Each of these traits is determined by a single gene. The results for five matings of specific plants (not necessarily true breeders) are shown in **Table 1.2**. Which alleles are dominant and which are recessive? What are the most probable genotypes for the parents in each cross?

Answer: The best approach is to consider each character separately. The square allele is dominant to the round allele, which is recessive. This can be seen in crosses 1, 4, and 5, where square crossed with square produces the phenotypic ratio three square to one round, so the square parents must be heterozygous. In the square-with-round cross (5), only square-flower progeny were produced; therefore, both parents were homozygous and all of the progeny are heterozygous, with square flowers. Red is dominant to white. This is easiest to see in crosses 1 and 3. For the genotypes, R = square, r = round, W = red, w = white. The best approach here is to consider each character separately and assign the recessive homozygous genotypes first, then decide which of the dominant parental phenotypes must be heterozygous. Cross 1 is $WWRr \times wwRr$; cross 2 is $Wwrr \times wwRr$; cross 3 is $WwRr \times Wwrr$; cross 4 is $WwRr \times wwRr$; cross 5 is $WwRR \times wwrr$.

Chapter Test

True/False

1. Mendel collected too many progeny for his analyses.
2. It is essential to know the parents of breeding organisms in order to interpret their progeny.
3. The mating of a homozygous dominant with a homozygous recessive organism produces only offspring with the dominant phenotype.
4. A true breeder is an individual that is homozygous for the recessive allele.
5. The recessive phenotype is the appearance masked in the F_1 of a cross between two dissimilar true breeders.
6. A genotype is the set of alleles possessed by an individual.
7. Alleles are bits of information that are passed, one from each parent, to progeny.
8. Independent assortment causes equal representation of both alleles in the gametes of an *Aa* organism.
9. The best method for deciphering genotypes is a cross of the unknown with a dominant homozygote.

Multiple Choice

10. The cross of a heterozygous yellow pea plant with another heterozygous yellow pea plant would most likely produce
    a. all yellow offspring.
    b. all green.
    c. 1,198 yellow and 1,212 green.
    d. 1,787 yellow and 596 green.
    e. 1,829 green and 605 yellow.

11. How many different kinds of gametes can a trihybrid produce (assume independent assortment)?
    a. 2
    b. 4
    c. 6
    d. 8
    e. 12

12. Which of the following features of meiosis is not correct?
    a. Four haploid cells are produced from one diploid cell.
    b. Recombination is common.
    c. The products are genetically identical cells.
    d. The products are gametes.
    e. Chromosomes are seen to separate according to Mendel's principles.

13. Which of the following is incorrect? A 1:2:1 ratio
    a. is a phenotypic ratio.
    b. is a genotypic ratio.
    c. represents the F_2 generation in Mendel's experiment.
    d. results from the selfing of a monohybrid.
    e. represents dominant homozygotes, heterozygotes, and recessive homozygotes, respectively.

14. Independent assortment ensures that an *AaBb* individual will produce gametes
    a. of one type.
    b. of two types, with the dominant types outnumbering the recessive types.
    c. of two types in equal proportions.
    d. of four types, with the parental types outnumbering the novel types.
    e. of four types in equal proportions.

15. The characteristic ratio 9:3:3:1 indicates all of the following except:
    a. The parents of these progeny were both double heterozygotes.
    b. The dominant alleles were more numerous than the recessives in the progeny.
    c. The alleles are clearly dominant and recessive.
    d. The alleles of each gene segregate equally.
    e. The genes in the cross assort independently of each other.

16. Sutton and Boveri postulated that
    a. genes are parts of chromosomes.
    b. two characters will be partitioned randomly into the progeny.
    c. gametes will represent equally all combinations of alleles present in the parent.

d. some alleles are dominant while others are recessive.
e. sex is chromosomally determined.

Short Answer

17. The cross of a yellow round pea to a green round pea produced 50 green wrinkled, 147 green round, 54 yellow wrinkled, and 160 yellow round offspring. How do you account for these numbers?

18. Raspberries can be red (*R*) or black (*r*) and fuzzy (*B*) or bare (*b*). If a *RrBb* plant is selfed, what fraction of the progeny will be black? fuzzy? black and fuzzy? red and fuzzy?

Essay

19. ompile a description of all of the ratios presented in this chapter along with all of the crosses that produce those ratios.

Chapter Test Answers

1. **F** 2. **F** 3. **T** 4. **F** 5. **T** 6. **T** 7. **T** 8. **F** 9. **F** 10. **d** 11. **d** 12. **c**

13. **a** 14. **e** 15. **b** 16. **a**

17. The color numbers, 197 green and 214 yellow, look like a testcross of the color character. The parents were heterozygous yellow and homozygous green. (We cannot tell from *this cross* which color is dominant.) The shape numbers, 104 wrinkled and 307 round, look like the results of a dihybrid cross in which round is the dominant trait. Parental genotypes were *GgWw* × *ggWw* [*G* is yellow (Mendel showed yellow to be dominant); *g* is green; *W* is round; *w* is wrinkled].

18. 1/4 black, 3/4 fuzzy, 3/16 fuzzy black, 9/16 fuzzy red

19. All dominant progeny come from three crosses: homozygous dominant with homozygous dominant, or with homozygous recessive, or with heterozygous. All recessive progeny come from homozygous recessive crossed to homozygous recessive. Monohybrid selfings produce 1:2:1 genotypic ratios and 3:1 phenotypic ratios. Crosses of heterozygotes to homozygous recessives produce 1:1 progeny. Dihybrid selfings produce progeny in a 9:3:3:1 phenotypic ratio. Homozygous dominant organisms crossed to any organism will yield all homozygous dominant progeny (a 1:0 ratio). Homozygous recessive selfings give all homozygous recessive progeny (a 0:1 ratio). Dihybrid crossed to homozygous recessive gives a phenotypic ratio of 1:1:1:1 progeny. There are also variations that give other ratios, such as 3:1:3:1 (also see the Demonstration Problem above).

Check Your Performance:

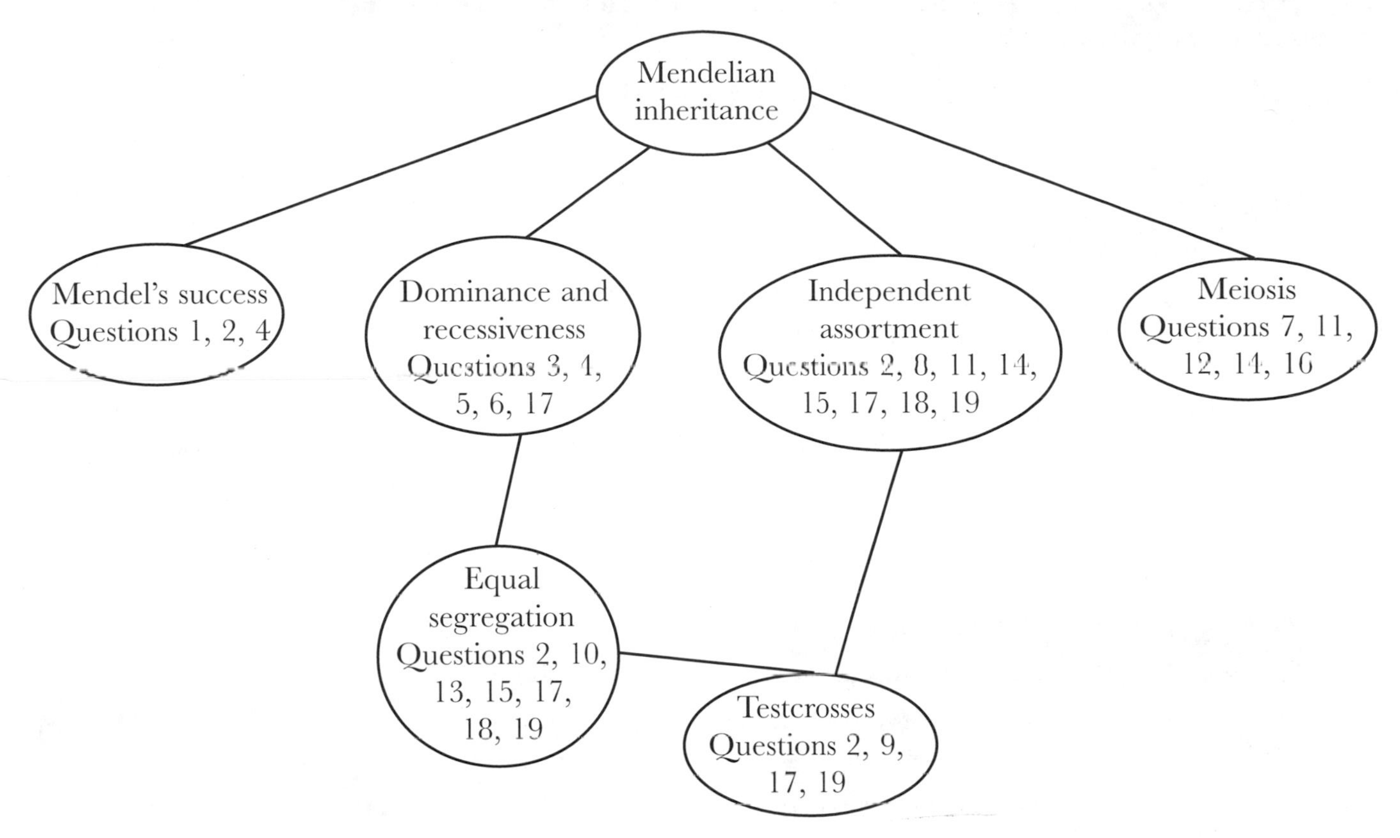

Use this chart to identify weak areas, based on the questions you answered incorrectly in the Chapter Test. This subject matter forms the basis of modern genetics and many more connecting lines could be drawn between these topics. If you feel weak in *any* area listed above, review it before going on.

Probability and Statistics

Where do genotypic ratios and phenotypic ratios come from? In the monohybrid cross $Aa \times Aa$, where A and a are two alleles of a generic gene, the ratios are obvious from the Punnett square, but they can also be calculated. For more complex crosses, trihybrids, for example, the Punnett square would be difficult to generate and to use for ratio determination. Instead, you will need to master a few tools for calculating probabilities. Some of these tools will enable you to determine the frequencies of specific classes of progeny without simultaneously determining all other classes. Let us begin with a definition of probability.

ESSENTIAL BACKGROUND

- **Recessiveness and dominance (Chapter 1)**
- **Principle of segregation (Chapter 1)**
- **Principle of independent assortment (Chapter 1)**
- **Gamete formation (Chapter 1)**
- **Punnett squares (Chapter 1)**

TOPIC 1: PROBABILITY AND THE PRODUCT RULE

KEY POINTS

✓ *What is probability?*

✓ *What is the product rule?*

✓ *How is the product rule used to calculate genotype and phenotype frequencies?*

Probability is the likelihood of a particular event happening. Put another way, it is the number of times an event has occurred divided by the number of times it could have occurred. This quantity has predictive value for future occurrences of the same event. Probabilities (p) are generally reported as fractions or decimals, ranging in value from 0 to 1. If the probability of an event is 0, it never happens; if the probability is 1, it is a sure thing. The classic examples of commonly encountered probabilities are coin tosses, dice rolling, and sex of babies. The probability of a tossed penny landing head-side up is 1/2, because there are two possible ways for the coin to land. If we rolled one die, the probability of getting a four, p(four), is 1/6 (there are six sides to each die). We have determined the probability of gametes in the same way. The monohybrid Aa makes gametes that are $p(A) = 1/2$ and $p(a) = 1/2$, meaning equal proportions of the two types of gametes are produced. The dihybrid $AaBb$, where B and b are two alleles of another generic gene that shows independent assortment from the *A* gene, gives four kinds of gametes at equal frequency: 1/4 *AB*, 1/4 *Ab*, 1/4 *aB*, 1/4 *ab*. Once the gametes and their frequencies have been determined, the next step is to decide the genotypes and frequencies. Genotype frequencies for the offspring are calculated using the product rule.

The **product rule** states that the probability of two (or more) independent (i.e., mutually exclusive) events occurring together is the product of the probabilities of each event occurring alone. (One hint for whether the product rule can be used to determine a probability is to ask whether the word *and* can be used to link the requisite events.) Think of two dice being thrown. As one comes to rest with the four-side up, it does *not* influence the other die to also land four-side up; hence, they are independent events. The product rule is applicable to exactly this kind of situation. The probability of one die landing with a particular number showing is 1/6. The probability of two dice rolling fours is the product of 1/6 and 1/6, that is, the product of the probability of each event occurring alone, or 1/36. (Using the "and" hint, four *and* four is what is needed.) Two other events that are independent but occurring together involve the generation of the gametes that make an offspring. Gamete formation is independent in the two parents, so the two gametes that unite to form a single offspring have independently determined genotypes. Their union to make the offspring's genotype is the "occurring together" part of the product rule. If one monohybrid produces an *A* gamete and its mate produces an *A* gamete, each have a probability of 1/2, so the probability of the offspring having the genotype *AA* is $1/2 \times 1/2$, or 1/4. That is why the *AA* progeny of the monohybrid cross amounted to 1/4 of the total progeny.

Topic Test 1: Probability and the Product Rule

True/False

1. The product rule can only be used when the desired events occur contemporaneously.
2. If 1 of every 10,000 babies born in a particular city has the recessive disease phenylketonuria, and if 1 affected baby were just born, another affected baby will not be born until 9,999 normal babies have been born.

Multiple Choice

3. What is the probability that a family with two children have two male children?
   a. 1/8
   b. 1/4
   c. 1/3
   d. 1/2
   e. 1
4. What is the probability that the selfing of a tetrahybrid (*AaBbCcDd*) will generate a fully recessive individual as the first offspring?
   a. 1
   b. 1/16
   c. 1/64
   d. 1/256
   e. 1/1,024

5. Phenylketonuria is a rare, recessive disease in humans. If Susan's father is affected by this disease and her mother is not, what is the probability that she is a heterozygote?
   a. 1
   b. 3/4
   c. 1/2
   d. 1/4
   e. 0

Short Answer

6. A couple was surprised to have an albino child (albinism is caused by a rare, autosomal recessive allele). The husband's brother wants to marry the wife's sister, but they are concerned about the probability of having albino children, too. What probability do you think they have? (Hint: Work backward to the grandparents first.)

Topic Test 1: Answers

1. **False.** There need only be one criterion for grouping the chances together: same time, same family, same day of the week, same city, same mothers' first names, etc.
2. **False.** This is essentially the same question as the previous one, yet it trips up many students. All of the subsequent babies are independent of the recent, affected one. Furthermore, the affected child is in the past; it is now a sure thing. Since it is known to be affected, it might help to think its p(disease) = 1.
3. **b.** The probability of having one male child is 1/2, so the probability of having two male children is $1/2 \times 1/2 = 1/4$. The product rule can be used because the criterion of occurring together is met by the children being in the same family.
4. **d.** The probability of one recessive genotype (e.g., *aa*) is 1/4, so the probability of four is $1/4 \times 1/4 \times 1/4 \times 1/4 = 1/256$. Another way to calculate it is as follows: The probability is the product of the probability of two fully recessive gametes = $(0.5 \times 0.5 \times 0.5 \times 0.5) \times (0.5 \times 0.5 \times 0.5 \times 0.5)$.
5. **a.** Her dad is homozygous recessive (*aa*). Since her mother is not affected and it is a rare disease, it is likely that she is *AA*. Therefore, Susan is *Aa*.
6. If there has not been albinism in their families before, their parents were most likely *AA* × *Aa*, in order to make the couple *Aa* × *Aa*. This means their siblings each have a 1/2 probability of being *Aa*, and if they are, the probability of having an albino child is 1/4. So $1/2 \times 1/2 \times 1/4 = 1/16$, which is the collective risk of the brother and sister both being heterozygotes and also having an affected child.

TOPIC 2: SUM RULE

KEY POINTS

✓ *What is the sum rule?*

✓ *When is it appropriate to use the sum rule?*

Let us take another look at an example in Chapter 1: the Punnett square showing the progeny produced from the selfing of a dihybrid (see Figure 1.3). Several of the genotypes appear more than once in the progeny boxes because they arise in different ways. For example, if we merely counted the instances of the genotype *GgWW*, we would determine its frequency to be 2/16. There is, however, an easier method for determining frequencies that does not require the knowledge of all other progeny, as this one did. Begin the alternative method by recognizing that the genotype *GgWW* arises from two different gamete combinations. One way to be *GgWW* is to inherit *GW* from the mother and *gW* from the father. A second, completely independent (i.e., mutually exclusive) means of getting the same genotype is to get *GW* from the father and *gW* from the mother. The probability of each of these two events is the product of p(mother's gamete) and p(father's gamete), or $1/4 \times 1/4 = 1/16$.

However, 1/16 represents only one combination, say, the mother's *GW* × father's *gW*. Since the combination of the father's *GW* × mother's *gW* yields the same progeny genotype, and it occurs at the same frequency, the total frequency is obtained by *adding* the probabilities of each independent event: p(mother's *GW*, father's *gW*) + p(father's *GW*, mother's *gW*) = $1/16 + 1/16 = 1/8$. This is the sum rule. The **sum rule** states that the probability of an event occurring, if there is more than one independent way for the event to occur, is the sum of all the probabilities of all the possible ways. The means must be independent, meaning they are mutually exclusive. (A hint for whether the sum rule can be used to determine a probability is to ask whether the word *or* can be used to link the requisite events—in the genotype example given, the mother's *GW* and father's *gW* *or* the father's *GW* and mother's *gW*.)

Families provide another good example of this rule. As we saw in Topic Test 1, the probability of a two-child family having two boys is $1/2 \times 1/2 = 1/4$, by the product rule. Now consider the probability of two-child families in which both kids are the same sex. There are now two, mutually exclusive kinds of families that satisfy this requirement, those with two boys and those with two girls. The p(two girls) is also 1/4. What is the probability that both children are the same sex (i.e., a boy and a boy or a girl and a girl)? Add together 1/4 (for two boys) and 1/4 (for two girls) = 1/2. So half of all two-child families should be expected to have same-sex children. The converse situation is the probability of two-child families whose children are not the same sex. One option is an older daughter and younger son. The second option has the son being older. The probability of each case is $1/2 \times 1/2 = 1/4$, so the probability of both is $1/4 + 1/4 = 1/2$. This makes sense, since the only other kinds of two-child families have same-sex kids and we already determined that 1/2 of two-child families are like this. This points to another fact: The sum of the probabilities of all the independent ways to satisfy a requirement, like two-child families, will be 1 (provided the only ways to get the outcome are independent): p(two children, same sex) + p(two children, not same sex) = $1/2 + 1/2 = 1$, where 1 is all (100%) of two-child families.

Topic Test 2: Sum Rule

True/False

1. The probabilities of independent events can be added together to get the likelihood that either will cause a particular outcome.
2. There is no limit on the number of events whose probabilities can be added together for the p(outcome) provided the events are independent.

Multiple Choice

3. In the jimsonweed, purple flowers are dominant to white. A particular heterozygous jimsonweed is self-fertilized and only two of the resultant seeds are planted. What is the probability that both will produce flowers of the same color?
   a. 1/16
   b. 1/4
   c. 1/2
   d. 9/16
   e. 5/8

4. What is the probability of rolling a sum of six on two dice?
   a. 1/36
   b. 3/36
   c. 5/36
   d. 1/6
   e. 5/12

Short Answer

5. Chin clefts are caused by a rare dominant allele. If two people, both of whom are heterozygous for the chin-cleft allele, had three children, what is the probability that all the children will have the same phenotype?

6. What is the probability that the people in question 5 will have at least one child without the chin-cleft phenotype?

Topic Test 2: Answers

1. **True.** This is the definition of the sum rule. The events must be independent of each other (i.e., mutually exclusive). Then their sum gives the likelihood that the outcome of either one will be observed.

2. **True.** There is no mathematical limit, although as the number of events increases, it becomes increasingly difficult to count them all.

3. **e.** The probability of purple flowers is 3/4 per plant, or $3/4 \times 3/4 = 9/16$ for two plants. The probability of white flowers is 1/4 per plant, or $1/4 \times 1/4 = 1/16$ for two plants. These are the two mutually exclusive ways to meet the question's requirement for same color, so add the two probabilities together, $9/16 + 1/16 = 10/16 = 5/8$.

4. **c.** There are five ways for two dice to sum to six: 5 and 1, 1 and 5, 4 and 2, 2 and 4, and 3 and 3. The odds of rolling each number is 1/6, so the probability for rolling each pair is $1/6 \times 1/6 = 1/36$. Each of the rolls is independent and mutually exclusive, so add 1/36 for each of the ways: 5/36.

5. The mating is $Cc \times Cc$, so the probability of chin-cleft offspring is 3/4 and smooth-chin offspring is 1/4. The probability of all children being affected is $3/4 \times 3/4 \times 3/4 = 27/64$ and all children being unaffected is given by $1/4 \times 1/4 \times 1/4 = 1/64$. The probability of all three children having similar chins is $27/64 + 1/64 = 28/64 = 7/16$.

6. The answer to this question is easier to determine by calculating the probability that all children will have the chin cleft (27/64) and subtracting that from 1. The probability that at least one child will not have a cleft chin = 1 – p(all children will have cleft chin) = 1 – 27/64 = 37/64.

TOPIC 3: CONDITIONAL PROBABILITY

KEY POINTS

✓ *What is conditional probability?*

✓ *When is it appropriate to use conditional probability?*

One additional kind of probability calculation is used when the outcome whose probability is desired is contingent upon a condition, hence, the name **conditional probability**. The idea behind this concept is that placing a condition on the outcome limits the number of other events that will have to be considered in the calculation. The formula is $P_c = p_a/p_b$, where p_a is the probability of the desired outcome and p_b is the probability of the condition being met. An especially graphic example of conditional probability is the "risk" calculation for heterozygosity in an unaffected individual whose family segregates a recessive disease allele. Consider the tragic Tay-Sachs disease, which causes death during early childhood in recessive homozygotes. If a man's sister died from this disease, he may wish to know his risk of carrying the recessive allele. First, we can conclude that both of his parents are heterozygotes: Neither can be homozygous recessive, and only if both are heterozygous can they have a recessive homozygous child. On the surface of this problem, it seems that his risk is 1/2, because that is the probability that any single offspring of a monohybrid cross is heterozygous. However, this probability considers that the other 1/2 of the offspring are the two types of homozygotes, but we know the man is *not* a recessive homozygote because he is alive. Therefore, we need to calculate a probability that considers the condition, which is not having the disease. In the equation, p_a is 1/2, because half of all progeny are heterozygous, and p_b is 3/4, because 3/4 of all progeny are unaffected. $P_c = (1/2)/(3/4) = 2/3$. The probability that the man is heterozygous is 2/3, because we can exclude one of the other possible genotypes.

Topic Test 3: Conditional Probability

True/False

1. The calculation of genetic disease risks is improved by using conditional probability rather than solely the product rule.
2. To estimate an unaffected person's risk of heterozygosity, we need to know how he or she is related to affected family members.

Multiple Choice

3. Phenylketonuria is a rare, recessive disease in humans. If Susan's brother is affected by this disease, and everyone else is unaffected, what is the probability that she is a heterozygote?
   a. 0

b. 1/4
c. 1/2
d. 2/3
e. 3/4

4. Larry's uncle had cystic fibrosis, an uncommon recessive disease. The other side of Larry's family has no history of the disease. What is the probability that Larry is heterozygous for this recessive allele?
a. 1/4
b. 1/3
c. 1/2
d. 2/3
e. 1

Short Answer

5. A man's aunt had phenylketonuria (his parents were phenotypically normal and one had no family history of the disease). His wife knows she is heterozygous for the phenylketonuria allele.
a. What is the probability that their first child will have this disease?
b. What is the probability that, assuming they have three children, all will be normal?

Topic Test 3: Answers

1. **True.** Risk calculation is made more accurate by counting heterozygote risk as 2/3 (via conditional probability) instead of 1/2 (via product rule) for an autosomal recessive disorder.

2. **True.** To calculate risks of this sort, we have to assign genotypes to the parents, so it is very important to know which relatives are affected.

3. **d.** The fact that Susan's brother has the disease tells us that their parents were probably heterozygotes, since they were unaffected. Susan is among the 3/4 of their progeny that are unaffected. One-half of all progeny are heterozygous, so the probability that Susan is heterozygous is $P_c = (1/2)/(3/4) = 2/3$.

4. **b.** The probability that Larry's parent is heterozygous is 2/3 and the probability that Larry would inherit this allele (assuming the parent is heterozygous) is 1/2. Both have to be true for Larry to be heterozygous, so use the product rule: $2/3 \times 1/2 = 1/3$.

5. **a.** First, calculate the probability that the man is heterozygous. Since the man's aunt was affected, both of her parents (his grandparents) must have been heterozygotes. The man's parent (sibling of the aunt) is normal, so his or her probability of being heterozygous is 2/3. This person's mate is very likely to be a dominant (unaffected) homozygote because this is a rare disease. If we assume that the man's parents' genotypes are *Pp* and *PP*, the probability that he is heterozygous is 1/2. So the probability that the man is heterozygous is the probability of his inheriting the recessive allele from the first parent, multiplied by the probability of that person having the allele (i.e., being heterozygous), or $1/2 \times 2/3 = 2/6 = 1/3$. If the man is heterozygous, his and his wife's genotypes are *Pp* and *Pp*, and so

there is a 1/4 chance that one child is affected (*pp*). The total probability of the child being affected is p(dad is heterozygous) × p(child is *pp*) = 2/6 × 1/4 = 2/24 = 1/12.

b. The probability that one child is normal is 1 – 1/12 = 11/12. The probability that all three children are normal is 11/12 × 11/12 × 11/12 = 1,331/1,728 = 0.77.

TOPIC 4: BINOMIAL PROBABILITY

KEY POINTS

✓ *How can you find the probability of a particular outcome if there are a large number of ways it can happen?*

✓ *What are the advantages to using the binomial probability?*

You now know how to combine the product rule and sum rule to determine probability when more than one way exists to reach a particular outcome. But it should be obvious that in many situations that fit this description, it will not be easy to determine probability with the product rule–sum rule method. For example, what proportion of seven-child families consist of four girls and three boys? This question is complicated because there are more than a few birth orders that satisfy the four-girl, three-boy criterion. The combined product rule–sum rule method runs the risk of missing some orders, since there are 35 possible orders. The method of choice in this case is to calculate a **binomial probability**. The name of this probability comes from the formula's basis in the binomial distribution. The formula is $(n!/s!\ t!)\ a^s b^t$, where n is the total number of events (= s + t), s is the number of "a" events, t is the number of "b" events, a is the probability of the "a" event occurring, and b is the probability of the "b" event occurring. Notice that the n, s, and t terms are factorials. Factorials are defined as n! = n(n – 1)(n – 2) (n – 3) . . . (1); for example, 4! = 4 × 3 × 2 × 1 = 24. (0! equals 1.) Another caveat is that a + b must equal 1; that is, the "a" event and the "b" event must be the only two possible outcomes. To answer the seven-child family question: n = 7, s(girls) = 4, t(boys) = 3, a(girl) = 1/2, b(boy) = 1/2. This gives $7!/4!3!\ (1/2)^4(1/2)^3 = 35\ (1/2)^7 = 35/128 = 0.27$. If you wanted to know the probability of seven-child families with at least four girls, you would also calculate the binomial probabilities for families of five, six, and seven girls, and then add the four probabilities together. This technique is used extensively in population genetics and for the analysis of quantitative traits as well as for figuring the probabilities of complex assortments of progeny.

Topic Test 4: Binomial Probability

True/False

1. When there are one or two mutually exclusive ways for an outcome to occur, the binomial probability can still be used to determine the probability of that outcome.

2. To use the binomial method, all of the independent ways of reaching the outcome do not need to be enumerated.

Multiple Choice

3. An individual who is heterozygous for the recessive albinism allele is mated to an individual who is albino (homozygous recessive). What is the probability that exactly two of their five children will be albino?

a. 5/96
b. 1/32
c. 1/4
d. 5/16
e. 1/2

4. If two people who are heterozygous for the ability to taste phenylthiocarbamide have five children, what is the probability that exactly three will be tasters (tasting is the dominant phenotype)?
a. 0.00264
b. 0.0879
c. 0.264
d. 0.313
e. 0.879

Short Answer

5. The first child born to a normally pigmented couple has the recessive disorder albinism. If the parents have six more children, what is the probability that four of the children will be normal and two will be albino?

Topic Test 4: Answers

1. **True.** Both the binomial formula and the product rule are equally valid approaches to this type of problem.
2. **True.** The binomial method automatically accounts for multiple possibilities without listing them.
3. **d.** The mating is *Aa* × *aa*, so each child has a 1/2 (= a) chance of being albino and a 1/2 (= b) chance of being normal, n = 5, s = 2, and t = 3. Using the formula, 5!/2!3! $(1/2)^2(1/2)^3$ = 10/32 = 5/16.
4. **c.** n = 5, s = 3, t = 2, a = 3/4, b = 1/4. The calculation is 5!/3!2! $(3/4)^3(1/4)^2$ = 0.264.
5. First, interpret the firstborn child as an indication of the parents' heterozygosity for the recessive allele. Then ignore that child, since her phenotype does not affect the phenotypes of the remaining children, and focus on the next six: 6!/4! 2! $(3/4)^4(1/4)^2$ = 0.297.

TOPIC 5: SIGNIFICANCE TESTING WITH THE CHI-SQUARE ANALYSIS

KEY POINTS

✓ *How do we tell whether our observed data match a predicted ratio?*

✓ *What are the necessary elements of chi-square analysis?*

✓ *How do we perform a chi-square analysis?*

✓ *How do we interpret the results of chi-square analysis?*

Table 2.1 Chi-Square Analysis

PHENOTYPE	o*	HYPO	e	o – e	$(o - e)^2$	$(o - e)^2/e$
yellow round	**315**	9/16	312.75	2.25	5.06	0.0162
yellow wrinkled	**108**	3/16	104.25	3.75	14.06	0.1349
green round	**101**	3/16	104.25	–3.25	10.56	0.1013
green wrinkled	**32**	1/16	34.75	–2.75	7.56	0.2176
Total	556		556			$\chi^2 = 0.47$

hypo = fraction predicted from the hypothesis; o – e = observed minus expected; χ^2 = sum of the last column.

* The numbers in bold are the actual numbers of progeny that Mendel counted in each class.

This topic addresses the critical question of what causes the differences between real numbers, such as the results of Mendel's crosses, and the values expected on the basis of a hypothesis. Is the difference due to chance variation or is it a real difference between the hypothesis and the results? Mendel counted hundreds or thousands of peas for each cross to minimize the effect of error on his conclusions, because as the numbers of each type of progeny increase, random fluctuations will make less difference between the real numbers and the ideal numbers. The problem for geneticists and students is to decide whether the data generated by an experimental mating fit a hypothetical progeny ratio, for example, 3:1. For this reason, the statistical analysis of data is called **testing for "goodness of fit"** and a commonly used test of this sort is the **chi-square method**. This test compares the numbers of progeny in each class to those predicted from a genetic hypothesis.

The results of Mendel's dihybrid cross make a good example of this test. The formula is $\chi^2 = \Sigma (o - e)^2/e$, where χ is the Greek letter chi (pronounced kie, to rhyme with pie), Σ is the sign for summing together the term that follows for each progeny class, o is each observed value, and e is each expected value, predicted from a genetic hypothesis. Some students find it easier to use and to remember a table, like **Table 2.1**, with separate columns for each part of the calculation.

When a chi-square analysis is performed, the hypothesis should always be clearly and specifically stated in words as well as numbers: The hypothesis being tested in Table 2.1 is segregation of two genes that specify simple dominant and recessive traits and that exhibit independent assortment, mathematically represented as 9:3:3:1. Expected numbers come from this ratio, using the total number of progeny observed (556), multiplied by the fraction of each class expected from the hypothesis (e.g., 9/16), to get numbers of the right size for the comparison. To test the fit of the observed numbers to the hypothesis, the expected ratio *must* be scaled up to produce the exact numbers of progeny that should have been observed in each class in the absence of variation. Note that fractions are allowed! Next, the difference between the observed and expected is determined, squared, and divided by the expected. Once this has been done for all four progeny classes, the deviations in the last column are added up to get the χ^2 value.

The interpretation of the χ^2 value is not complete until it has been analyzed along with the number of classes in the experiment. This is done with a parameter called **degrees of freedom**, or **df**. In genetic usage of chi-square analysis, one degree of freedom is allowed for all but one of the progeny classes in the analysis (df = n – 1, where n is the number of classes). For Mendel's dihybrid cross, there were four progeny classes (n = 4), so df = 4 – 1 = 3. This parameter allows the experiment to show an amount of deviation proportional to the number of

Table 2.2 Chi-Square Probabilities

	p VALUE				
df	**0.9**	**0.5**	**0.1**	**0.05**	**0.01**
1	0.016	0.455	2.706	3.841	6.635
2	0.211	1.386	4.605	5.991	9.210
3	0.584	2.366	6.251	7.815	11.345
4	1.064	3.357	7.779	9.488	13.277

progeny classes observed, by scaling the χ^2 value for the number of classes. Finally, the χ^2 and df values are used to calculate a probability, p, that tells us the probability that these results would be obtained, if the hypothesis is correct. If the p value is large, then it is very likely that the hypothesis fits these data, but if the p value is small, then the results are not very likely to be the result of the hypothesis tested. Probability is normally calculated by a computer program but it is easier for students to use a chart. An abbreviated form of this chart is in **Table 2.2**. Values for χ^2 are within the chart, those for df are down the left side, and p values are across the top. To use the chart, find the correct row for your df value, find the right χ^2 value in that row, or a pair of values that bracket yours, then trace up to the top of the column to get the p value. If p is greater than 0.05, the hypothesis is accepted, which is to say that an experimental test of this hypothesis is expected to show this much or greater deviation solely as a result of chance more than 5% of the time. The greater the p value, the more likely the hypothesis is correct. Values of p that are less than 5% (0.05) generally cause the hypothesis to be rejected. This does not mean that the hypothesis is disproved, only that it is *unlikely* to explain these data. In the example of Mendel's dihybrid cross, $\chi^2 = 0.47$ and df = 3, so $p > 0.9$. We conclude that the hypothesis can be accepted.

If we had been forced to reject the hypothesis, there are two causes to consider: (1) The size of the progeny classes may be too small, increasing the likelihood that they could show extreme variation despite the hypothesis being correct, or (2) some aspect of the hypothesis is wrong. What do we do if our p value is less than 0.05? We should either repeat the analysis with a larger sample size and see whether this increases the p value, or test a new hypothesis. Chi-square analysis can be used whenever the hypothesis can be stated mathematically, and is especially useful for distinguishing between alternative hypotheses. You already know ratios that reflect alternative hypotheses; four examples are 3:1, 1:1, 9:3:3:1, and 1:1:1:1. For any given data set testing various ratios by chi-square analysis will reveal the best hypothesis as the one having the largest p value. You will learn more ratios in Chapter 3.

Topic Test 5: Significance Testing with the Chi-Square Analysis

True/False

1. The chi-square test begins with a statement describing what hypothesis is being tested.

2. The χ^2 value provides enough information to determine acceptance or rejection of hypothesis.

3. If the p value is very small, there is little chance the tested hypothesis is correct.

Multiple Choice

4. To perform a chi-square analysis on the results of a trihybrid cross, how many degrees of freedom are appropriate?
   a. 1
   b. 3
   c. 7
   d. 8
   e. 16

5. How is the hypothesis ratio compared to the observed numbers in a chi-square test?
   a. Divide the observed numbers by the total of the expected ratio and round off to whole numbers.
   b. Divide the observed numbers by the total of the expected ratio, allowing fractions.
   c. Compare the expected ratio to the observed numbers just as they are.
   d. Multiply the expected fractions by the total number of progeny and round off to whole numbers.
   e. Multiply the expected fractions by the total number of progeny and allow fractions.

Short Answer

6. Each of the kernels in an ear of corn is the progeny of a mating between two corn plants. For the data below, assume that all of the kernels come from the same mating. Kernels can be smooth or wrinkled and they can also be purple or yellow. If a dihybrid cross produces the kernels shown below, what can you conclude about these traits? State your hypothesis clearly and test it with a chi-square test.

phenotype	number observed
Yellow smooth	499
Yellow wrinkled	151
Purple smooth	156
Purple wrinkled	62

Topic Test 5: Answers

1. **True.** The hypothesis that is being tested must be stated in order to calculate correctly the expected values for testing the hypothesis.

2. **False.** The χ^2 value measures *total* deviation and therefore is insufficient to know whether to accept or reject the hypothesis. Degrees of freedom must also be taken into account.

3. **True.** A low p value means that the difference between what was observed and what was expected is so large that it is unlikely to be seen in a test of the same size where this hypothesis is true.

4. **c.** The ratio that will be tested is 27:9:9:9:3:3:3:1, which is eight phenotypic classes. The convention is to use one fewer degree of freedom than there are classes.

5. **e.** A common mistake students make in the chi-square test is to use the wrong values for observed and expected. Do *not* change the observed values. *Do* change the hypothesized ratio, to make it sum to the same size sample as the observed numbers, but distributed in accordance with the hypothesized ratio, allowing fractions.

6. The hypothesis that should be tested includes the following provisions: Yellow and smooth are dominant to purple and wrinkled, respectively; these are segregating alleles of two independently assorting genes. This hypothesis leads to the expected ratio 9:3:3:1. The expected numbers of progeny are 488.25 yellow smooth, 162.75 yellow wrinkled, 162.75 purple round, and 54.25 purple wrinkled. $\chi^2 = (499 - 488.25)^2/488.25 + (156 - 162.75)^2/162.75 + (151 - 162.75)^2/162.75 + (62 - 54.25)^2/54.25 = 0.237 + 0.280 + 0.848 + 1.11 = 2.48$. There are 3 df, which gives a p value >0.5, so we accept the hypothesis. The data are consistent with the two independently assorting genes, with the yellow and smooth alleles being dominant to purple and wrinkled, respectively, which are recessive.

IN THE CLINIC

Probabilities and statistics are used extensively in genetic counseling. Counselors use family histories, laboratory tests, or both to determine the likelihood that a person has an affected genotype or is a carrier (heterozygous) for a recessive disorder. Approximately 4% of newborns have a significant abnormality (about half of these are genetic) that is present at birth or recognized within the first year of life. Parents of such babies, or couples who suspect they might have genetic disorders in their families, seek genetic counseling to determine the likelihood of having additional affected children, or a first child with the disorder. Counselors are trained in genetics, medicine, and counseling and they provide individuals and families with support services as well as information on the diseases, treatments, and options.

Chapter Test

True/False

1. Probability is the likelihood something will happen in the future.
2. The sum rule requires that the probabilities for getting an outcome are mutually exclusive.
3. The product rule is the best way to determine the probability of drawing a full house in poker (three of one value of card and two of another value of card) in the first five cards dealt from a full deck.
4. Binomial probabilities do not account for all possible orders of events.
5. Chi-square testing will indicate samples that are too small.
6. Observed measurements that are different from the expected values may not be significantly different and this hypothesis can be tested.
7. Smaller numbers are inherently more accurate.
8. As the number of classes increases, the total deviation will increase.

Multiple Choice

9. In the cross of a female with genotype *AabbCcDdEe* to a male of genotype *AaBbccDdEE*, what proportion of progeny will have exactly the same phenotype as the male parent?
   a. 1/32
   b. 9/64
   c. 9/32
   d. 3/16
   e. 1/2

10. How many degrees of freedom are allowed for the progeny of a pentahybrid cross that produces 32 progeny classes?
    a. 2
    b. 3
    c. 31
    d. 32
    e. None of the above

11. Phenylketonuria (PKU) is a metabolic defect caused by a recessive allele. If both parents are heterozygous for the disease allele, what is the probability that their normal child is heterozygous?
    a. 0
    b. 1/2
    c. 2/3
    d. 3/4
    e. 1

12. What is the probability that parents who are heterozygous for the recessive albinism allele will have four affected children?
    a. 0
    b. 1/1,024
    c. 1/512
    d. 1/256
    e. 1/16

13. What is the probability that parents who are both heterozygous for the recessive albinism allele will have four affected children out of six total?
    a. 0.00015
    b. 0.0039
    c. 0.0022
    d. 0.033
    e. 0.297

14. If a normal couple had four affected and two normal children, what would be the probability their next child will be affected?
    a. 0
    b. 0.00015
    c. 0.0022
    d. 0.012
    e. 0.25

15. What is the probability that two thrown dice will sum to 10?
    a. 1/6
    b. 3/36
    c. 5/36
    d. 3/6
    e. None of the above

16. The threshold probability for acceptance of the hypothesis in the chi-square test is
    a. 0.
    b. 0.01.
    c. 0.05.
    d. 0.10.
    e. 0.50.

Short Answer

17. A man has hairy palms, a rare dominant trait. Both he and his wife have another rare dominant trait, peg-shaped teeth, although each had one parent with normal teeth. What is the probability that of their two children, one has both traits and one has neither trait?

18. If the couple in question 21 had three children, what is the probability that at least one of them will have hairy palms?

Essay

19. Use the probability tools learned in this chapter to explain the monohybrid genotypic and phenotypic ratios.

20. Describe the use of the chi-square test.

Chapter Test Answers

1. **T** 2. **T** 3. **F** 4. **F** 5. **F** 6. **T** 7. **F** 8. **T** 9. **b** 10. **c** 11. **c** 12. **d**

13. **d** 14. **e** 15. **b** 16. **c**

17. 3/32

18. 7/8

19. The p(*AA*) is determined by the product rule: p(*A*) = 1/2 for each parent, so p(*AA*) = 1/2 × 1/2 = 1/4. The p(*aa*) is figured in the same way. The product rule also determines the probability of each way to get a heterozygote, that is, p(*Aa*), but then we must use the sum rule to account for both ways to get the heterozygote (*A* from the father versus *A* from the mother). Finally, to combine genotype frequencies for the phenotypic ratio, we use the sum rule again to add the 1/4 *AA* to 1/2 *Aa*. The result is 3/4 *A*-. Once more the sum rule gives p(*aa*) + p(*A*-) = 1; rearrange to get p(*aa*) = 1 – p(*A*-) = 1 – 3/4 = 1/4; or we could have used the product rule with gamete frequencies as for the genotypic ratio, above.

20. Chi-square analysis is used to test the conformity of a data set to a genetic hypothesis. The hypothesis generates a testable ratio, which is scaled up to the size of the data set. Each component of the data set is compared to the appropriate predicted value, and the

deviation between the two is determined. The deviations are squared and divided by the appropriate predicted value. These are summed and used with a counter of allowable deviation (chance deviation allowed in the experiment), called degrees of freedom. Together the summed deviations (the χ^2 value) and the degrees of freedom are used to determine the p value, the probability that this data set would be generated by any equivalently sized, experimental test of this hypothesis. Probabilities less than 0.05 are generally regarded as recommending rejection of the hypothesis.

Check Your Performance:

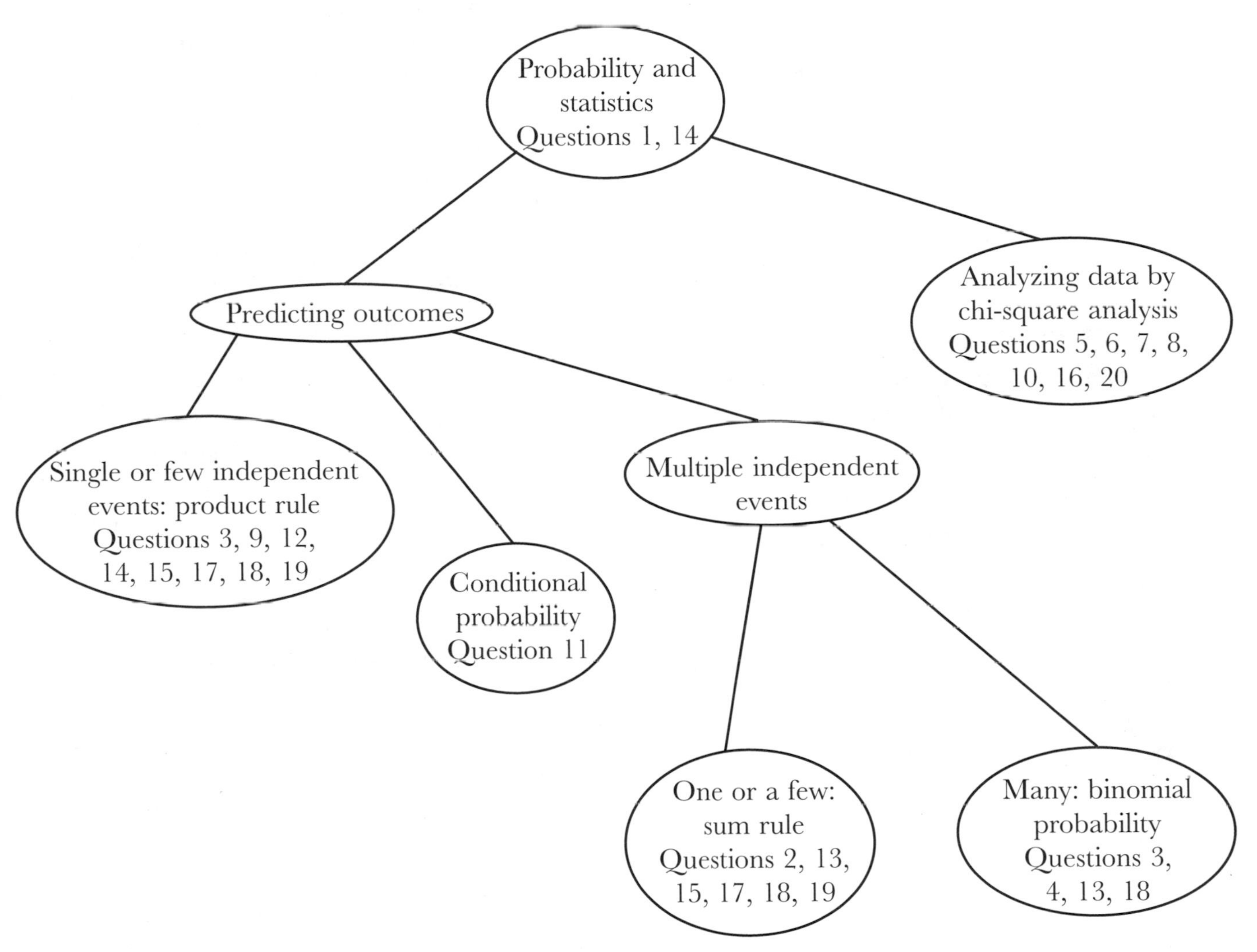

Use this chart to identify weak areas, based on the questions you answered incorrectly in the Chapter Test.

Extensions of Mendelian Inheritance

Mendel's observations form the basis of modern genetics. Dominance and recessiveness and the principles of random segregation and independent assortment are the starting points for understanding all genetic phenomena. These tenets are often true of newly discovered genes, but occasionally a gene is observed not to conform with one or more of the principles. This chapter concerns these special situations, providing examples of traits that violate the assumptions that all genotypes are observed as living, phenotypically simple organisms. Some traits exhibit more complex forms of dominance, and others have inviable genotypes. First though, we start with the manner in which Mendelian traits are detected in humans, because it involves a unique way to collect and display data on matings.

ESSENTIAL BACKGROUND

- **Segregation (Chapter 1)**
- **Dominance and recessiveness (Chapter 1)**
- **Independent assortment (Chapter 1)**
- **Mendelian ratios (Chapter 1)**
- **Punnett squares (Chapter 1)**
- **Gamete formation (Chapter 1)**

TOPIC 1: PEDIGREES

KEY POINTS

✓ *What is different about the study of human genetics?*

✓ *How are pedigrees depicted and interpreted?*

✓ *Does it matter whether a trait is rare or common?*

✓ *What are some common complications?*

In genetics, a **progressive study** is one where dissimilar organisms are crossed and their progeny are analyzed to determine what can be concluded about the inheritance of the traits. The progeny may be crossed among themselves or testcrossed or **backcrossed** to one of the parents. Geneticists who study human traits use **regressive studies**, to attempt to identify many informative matings that happened fortuitously, and then pool the matings and see if they suggest a mode of inheritance for the trait. Because of the small numbers of individuals in a typical

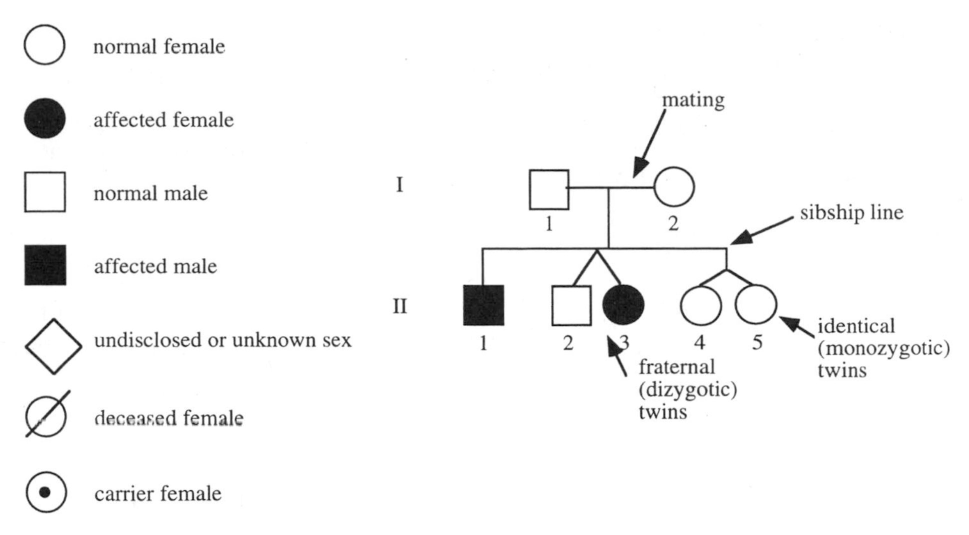

Figure 3.1 Symbols and sample of a pedigree.

family, often the interpretation will be ambiguous, but with large numbers of families a reasonable hypothesis is often possible. As you know from Chapter 2, small samples often show large, random variation from theoretical ratios as a result of chance. Consequently, although there are many Mendelian traits in humans (i.e., traits that are consistent with Mendel's principles), Mendelian ratios are rarely observed because the progeny numbers for single matings are tiny.

The data from families that show the trait are collected into family trees in a visual form called **pedigrees**. Although it is possible to depict several traits in a single pedigree, it is more common for pedigrees to follow only one trait. Some of the standard symbols for pedigrees, along with a sample pedigree, are shown in **Figure 3.1**. Circles represent females and squares represent males. The trait is indicated by shading the affected person's symbol. The member of the family who first came to the geneticist's attention, called the **proband** or **propositus**, can be indicated with an arrow. Matings are indicated by horizontal lines between symbols and the resulting progeny are connected to this line and to their siblings by a particular arrangement of vertical and horizontal lines (see example) that place the progeny symbols level with each other but lower than the symbols representing their parents, with the oldest on the left and youngest on the right. (All siblings' symbols for one pair of parents hang down from the same horizontal "sibship" line.) The convention is to name the rows of symbols with Roman numerals; for example, generation I is the first row and generation II is the second row. Each individual is then named by the appropriate Roman numeral plus an Arabic numeral denoting the person's position within the row, or generation. In Figure 3.1, the affected people (those who have the trait) are II-1 and II-3.

The task of interpreting pedigrees is made easier by a few guidelines. First let us tackle the subtle issue of common versus rare traits and alleles, because it is the most likely nuance to trick the novice. If a trait is caused by a recessive allele (so affected individuals are homozygous), and the trait is common, then many of the unaffected individuals in a pedigree are likely to be heterozygous. Alternatively, if the trait is rare, then the unaffected are likely to be homozygous dominant. Similar distinctions can be made for dominant traits. If the dominant allele that causes the trait is common, affected individuals are more likely to be homozygous than if the trait is rare.

Another set of guidelines is helpful for determining whether the trait is caused by a dominant versus recessive allele. If the trait is caused by a rare, recessive allele, affected offspring will have unaffected parents and the trait may be seen to skip generations. If the trait is common, one of the parents may also be affected. If the trait is dominant, it will not skip generations; affected parents will not always have affected offspring, but affected offspring will always have an affected parent. If the dominant trait is rare, affected people will be heterozygotes, and if it is common they are increasingly likely to be homozygotes.

Because sex in humans is determined by differing chromosomal constitutions, XX is female and XY is male, genes on these chromosomes are unequally represented in males and females. Hence, another clue to watch for in pedigrees is the presence of a recessive trait almost exclusively in males, rather than equally in both sexes. Such traits are said to be **X-linked recessive** for reasons that will become clearer in Topic 6. These affected males usually have normal children, but their daughters may have affected sons. **X-linked dominant traits** are recognized by affected males having affected daughters and unaffected sons. Both sexes of offspring of affected females can be affected. Traits that occur equally in both sexes are said to be **autosomal**; autosomes are the chromosomes that are not sex chromosomes (i.e., most chromosomes are autosomes). When you have a hypothesis, a good way to check it is by trying to assign genotypes to all individuals in the pedigree. Potentially this will allow you to catch inconsistencies between the hypothesis and the data. And remember to consider whether the trait is common or rare. (You will not always know which it is.)

Two additional complications in analyzing real pedigrees are worth mentioning: **penetrance** and **expressivity**. These are also the first examples of genetic phenomena that apparently are not consistent with Mendel's principles. If a trait is incompletely penetrant, not everyone with the "affected" genotype will show the trait. One human trait that is incompletely penetrant is polydactyly. Affected individuals are heterozygous for a dominant allele and have extra fingers or toes, or both. Most of the people who are heterozygous have the phenotype but some do not; these appear completely normal and are not identified as heterozygotes until they have affected children.

Traits that show variable expressivity do not cause exactly the same phenotype in all the individuals that have the "affected" genotype. Imagine a beagle; the spots on these dogs are caused by a variably expressive dominant allele. The resulting range of phenotypes are grouped by breeders into 10 classes, but genotypically they all are heterozygous or homozygous for the same dominant allele. If the variably expressive trait is fully penetrant, as it is for beagle spots, all "genotypically affected" individuals will have the affected phenotype. However, the range of "affected" phenotypes may be huge or small depending on the trait; there is no way to predict. Likewise, there is no way to predict how incompletely penetrant a trait is. This can only be determined from pedigree analysis, where an apparently unaffected person will be determined to have the affected genotype when the trait appears in his or her offspring or descendants. Incomplete penetrance and variable expressivity can be caused by environmental factors, genotypic variation of other genes, or a combination of both genetic and environmental factors.

Topic Test 1: Pedigrees

True/False

1. For a common recessive trait, many unaffected people will be heterozygous.
2. If a dominant trait is rare, the affected people are likely to be heterozygotes.
3. X-linked recessive traits are seen primarily in females.

Multiple Choice

4. What are the most likely observations for a common autosomal recessive disorder?
   a. Normal parents, all children affected
   b. Affected parent(s), some affected children
   c. Normal parents, some affected children
   d. Affected parent(s), no affected children

5. The following pedigree is inconsistent with which of the following modes of inheritance?

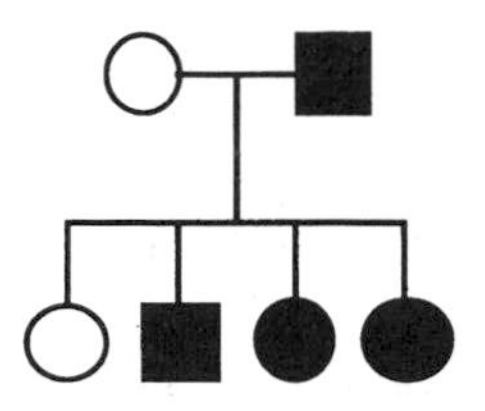

   a. Common autosomal recessive
   b. Common autosomal dominant
   c. Common X-linked recessive
   d. Rare autosomal recessive
   e. More than one of the above

6. Among people who have the autosomal recessive disease cystic fibrosis, some are severely ill while others have a much milder form of the disease. To what do you attribute this observation?
   a. Variable expressivity
   b. Incomplete penetrance
   c. Dominance and recessiveness
   d. Random segregation

Short Answer

7. Jane has a husband, a son, three younger sisters, two parents, and four grandparents. Many of the people in her family have a common dominant trait, attached earlobes, but her husband does not. Those whose earlobes are attached are Jane, her youngest sister, her father, both of her father's parents, and her maternal grandfather. Draw this pedigree and assign genotypes to all individuals.

8. Decide what mode of inheritance best explains the large pedigree below and cite which individuals aided your decision. (Assume it is a rare trait.)

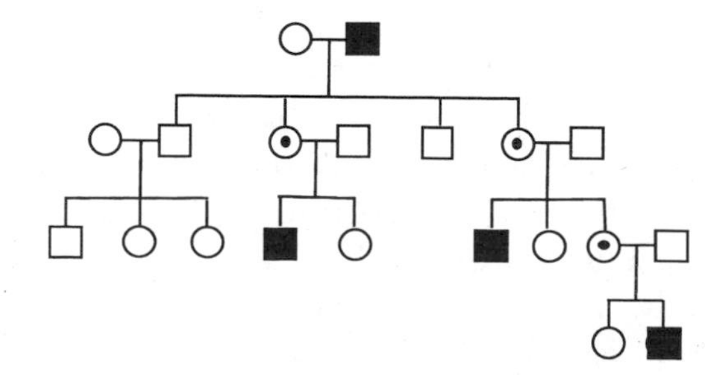

Topic Test 1: Answers

1. **True.** If the trait is common, then there must be many heterozygotes randomly mating to create these common homozygotes.

2. **True.** If the trait were common, there would be many affected heterozygotes mating to produce affected homozygotes. Since most matings are affected individuals to unaffected individuals, most of the affected progeny will be heterozygous.

3. **False.** X-linked recessive traits are primarily seen in males for reasons that will be clearer in Topic 6.

4. **c.** The terms *common* and *recessive* mean the normal parents have a good chance of being heterozygotes, and therefore would have some homozygous recessive progeny.

5. **d.** The least likely explanation is rare autosomal recessive because individual I-1 (the mother) would have to be heterozygous, which is unlikely if the trait is rare.

6. **a.** Traits that exhibit a range of phenotypes in all people of the same genotype have variable expressivity.

7.

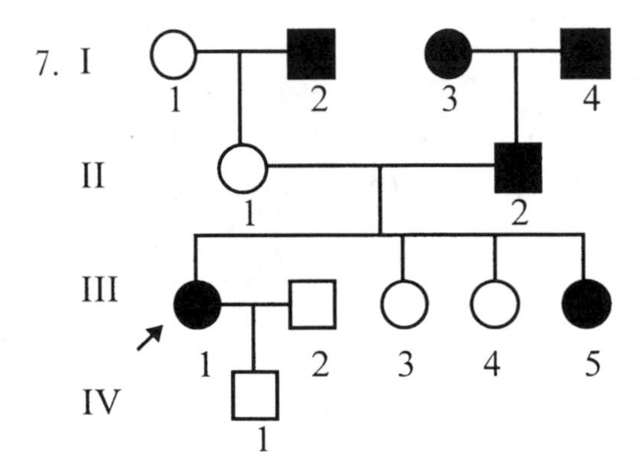

A is the dominant *attached* allele. *a* is the recessive *not attached* allele.

I-1 is *aa*. I-2 is *Aa*.

I-3 and I-4 are *A*- (either both are heterozygous or one is homozygous and one is heterozygous).

II-1 is *aa*. II-2 is *Aa*.

III-1 is *Aa*. III-2, III-3, and III-4 are *aa*. III-5 is *Aa*.

IV-1 is *aa*.

8. This trait appears to be X-linked recessive. The evidence is the existence of an affected male (I-2) who has normal offspring but affected grandsons; that is, the trait seems confined to the males, who are related by unaffected females. These affected grandsons (III-4 and III-6) are related to I-2 through his normal daughters, not his sons, consistent with X-linkage. The trait is not dominant, since it seems to skip generations, so it must be recessive. Finally, we cannot tell whether this trait is rare or common. We might guess rare, but without certainty.

TOPIC 2: INCOMPLETE DOMINANCE AND CODOMINANCE

KEY POINTS

- ✓ *What is incomplete dominance and how is it recognized?*
- ✓ *What is codominance and how is it recognized?*
- ✓ *How do these differ from Mendel's observations?*

This topic concerns two complications that Mendel did not observe. For him, particular alleles were always either dominant or recessive. The traits described here extend this concept and produce altered phenotypic ratios. But be aware that random segregation and independent assortment are still operating principles.

The first example is that of flower color in snapdragons, an ornamental plant whose blooms can be many different colors. If a true-breeding red snapdragon is crossed with a true-breeding white snapdragon, all the F_1 offspring will have pink flowers. This is the first sign that the flower-color character is not a simple Mendelian trait: The F_1 phenotype is intermediate to the two true-breeding phenotypes. If the F_1 offspring are selfed, the F_2 snapdragons appear in the ratio of 1 red:2 pink:1 white. This is the second clue: The F_2 ratio is not Mendel's F_2 phenotypic ratio; it is his F_2 genotypic ratio. Match the genotypes with the observed phenotypic classes: 1/4 are red and 1/4 are white; these are the homozygotes. One-half are pink, like the F_1; these are the heterozygotes. The Mendelian genotypic ratio is converted into this phenotypic ratio because the two alleles involved in the cross do not exhibit full dominance or recessiveness with respect to each other. We say they are **incompletely dominant** and we give them both capital symbols to indicate the absence of dominance. R^1 allele specifies red and the R^2 allele specifies white so the genotypes for the plants are R^1R^1 for red, R^1R^2 for pink, and R^2R^2 for white. Note that this relationship does not affect the inheritance of the gene, but it does reflect the function of the gene products. The R gene encodes an enzyme that normally catalyzes the synthesis of a red pigment. The R^1 allele encodes active enzyme, which makes the red pigment. The R^2 allele encodes an inactive enzyme, which is incapable of making the red pigment. The heterozygote is pink because the R^1 gene product makes red pigment but the R^2 gene product cannot make pigment. Half as much pigment in the heterozygote means pink flowers. To recap, incomplete dominance changes the 3:1 monohybrid phenotypic ratio to 1:2:1, with the phenotype of the heterozygous class intermediate to the homozygous phenotypes.

Codominance is closely related to incomplete dominance. Often the difference is merely how the phenotype is detected. One example of codominance is the MN blood group in humans. Normal red blood cells are coated in proteins, some with attached sugars. These proteins and sugars are specified by genes and vary from person to person. One such gene is responsible for the human MN blood groups, which are an example of codominant alleles. The mating of a phenotype M person to a phenotype N person produces only phenotype MN progeny. The key feature of **codominance** is that the heterozygote has *both* of the phenotypes of the two homozygotes.

A mating between two MN people will produce progeny in the ratio 1M:2MN:1N (with variation allowed for very small numbers). Type M and type N people are homozygous for the L gene; their genotypes are L^ML^M and L^NL^N, respectively. These symbols are designed to imply the lack of a clear dominance relationship. The non-Mendelian behavior of these alleles is a result of gene function. The protein specified by the L^M allele creates one moiety on the cell surface and the protein specified by the L^N allele creates a different moiety. Homozygotes, therefore, have one and not the other moiety. Heterozygotes have both because they have both gene products. Equal segregation of these alleles and independent assortment from other genes are still observed. The clues to codominance are the observed monohybrid cross ratio (three phenotypic classes in the ratio 1:2:1) and the presence of both alleles' phenotypes in the heterozygote.

Topic Test 2: Incomplete Dominance and Codominance

True/False

1. The genotypic ratio for a codominant trait is the same as for a Mendelian trait.
2. Incomplete dominance is caused by violation of the equal segregation principle.

Multiple Choice

3. A particular kind of summer squash comes in one of three colors (green, yellow, or white) and one of three shapes (long, oval, or round). The offspring of a cross between long green squash and round white squash are oval and yellow. When these progeny are crossed among themselves, they produce nine different kinds of offspring, in an interesting ratio. There are equal numbers of long green, long white, round green, and round white plants; twice as many long yellow, oval green, oval white, and round yellow; and four times as many oval yellow. Which of the following statements is most likely to be correct?
   a. Color and shape are two independently assorting genes whose alleles are dominant and recessive.
   b. Color and shape are two independently assorting genes whose alleles are incompletely dominant.
   c. Color and shape are determined by a single gene that has multiple alleles.
   d. These traits are incompletely penetrant.
   e. The long green and round white parents were not true breeders.
4. One breed of cattle can be red, white, or roan (a mixture of red and white hairs). The cross of two roans produces equal numbers of red and white progeny and twice as many roans. If a farmer wanted to breed an all-roan herd, what animals should be the parents?
   a. Roan × roan
   b. Red × red
   c. White × white
   d. Red × white
   e. Roan × red

Short Answer

5. What observation would cause you to conclude that a trait is codominant?
6. Regarding the previous question, could you devise an informative testcross? Explain.

Topic Test 2: Answers

1. **True.** Both are 1:2:1. It is the phenotypic ratios that differ.

2. **False.** Both incomplete dominance and codominance are inconsistent with the concepts of full dominance and recessiveness.

3. **b.** The appearance of new (intermediate) phenotypes in the F_1 offspring suggests incomplete dominance of both shape and color alleles. This hypothesis is confirmed in the F_2 generation by 1:2:1 ratios for long to oval to round and green to yellow to white progeny. NOTE: When you are confronted with a problem that seems to involve separate sets of traits, as this question does, try analyzing them separately.

4. **d.** Crosses a, d, and e all produce roan (heterozygous) offspring, but cross d produces only roans.

5. Two dissimilar phenotypes possessed by true breeders, and one phenotype that looks like the combined phenotypes of the true breeders but is not itself true breeding. Instead, crosses between the non-true breeders give progeny in a 1:2:1 ratio, the combined phenotype being the most numerous.

6. If the trait is codominant, then there is no recessive homozygote, so a testcross is meaningless. Crosses of the heterozygote to either homozygote yield both parental phenotypes in equal proportions. Crosses between the homozygotes give only the heterozygote.

TOPIC 3: LETHAL ALLELES

KEY POINTS

✓ *What is a lethal allele?*

✓ *When is it lethal? Why is it lethal?*

✓ *How can lethal alleles be recognized?*

Many gene products are absolutely essential for life. Lack or insufficient amounts of these proteins result in death, either sooner (as in embryogenesis) or later (as in early adulthood). A lethal allele is one that causes inviability either recessively or dominantly. For an example, consider the case of the Manx breed of cats. These cats have no tails, and a cross between any two Manx cats produces kittens in a surprising phenotypic ratio of 1 tailed cat to 2 tailless cats. Manx cats are never true breeders, which means they all must be heterozygotes. Call the alleles M^L and M, where M^L is the tailless allele. We know that M^L is dominant because the heterozygotes are tailless. Since the cross is between heterozygotes, we expect a 3:1 or 1:2:1 ratio for the offspring, but instead we see 1:2. This ratio is most similar to the 1:2:1 genotype ratio that Mendel observed and that we have seen adapted for use as a phenotypic ratio when there is no full dominance. However, it is not exactly the same ratio; one of the homozygous classes is missing.

The missing class cannot be the MM class because they have tails (they are the recessive homozygotes), so it must be the M^LM^L class that is missing. Why are they missing? Animals of this genotype die during gestation from acutely abnormal development. Since they are not seen in the progeny, the observed phenotypic ratio is 1:2. This allele has two phenotypes: taillessness, which is dominant, and lethality, which is recessive. This situation emphasizes the point that the only way to determine dominance and recessiveness is to see what heterozygotes look like. M^LM heterozygotes are tailless but not dead, and only M^LM^L homozygotes die.

Manx cats are an example of one type of **recessive lethality**. In this situation, one wild-type allele (M) is sufficient to allow the animal to survive, but the animal will not have a completely wild-type phenotype unless it has two wild-type alleles. In the other type of recessive lethality, one wild-type allele is sufficient to make the wild-type phenotype. Most lethal alleles are like this. Cystic fibrosis and muscular dystrophy are two examples of this type of lethality. Heterozygotes for these two genes have the wild-type phenotype, but homozygotes die prematurely.

Alleles can also be dominantly lethal. There are two mechanisms for **dominant lethality**. In one, the amount of the gene product is the critical factor in determining phenotype. This can be for one of two reasons: Either a single wild-type allele is not enough for a wild-type phenotype (**haplo-insufficiency**), or the non-wild-type (mutant) gene product interferes with a normal process within the cell or organism. The curious aspect to dominant lethality is that it is dominant. After all, how can lethality be dominant and still be inherited? Wouldn't the affected heterozygotes and homozygotes die before they could have progeny? If lethality occurred before childbearing, that would be true. But heritable dominantly lethal alleles cause death after the onset of childbearing ages. Huntington disease (HD) is a prime example. The average age of onset for HD is 39 years, and most people have had children by this age. This allele is very rare (*H*, because it is dominant) so most matings are $Hh \times hh$, where *h* is wild type. Progeny are expected in the ratio of 1 normal (*hh*) to 1 disease (*Hh*). That is, each child has a 50% chance of inheriting the disease-causing allele. In conclusion, dominantly lethal alleles are fatal after reproductive maturity is reached and recessively lethal alleles may be recognized by modified monohybrid ratios of 1:2.

Topic Test 3: Lethal Alleles

True/False

1. Recessive lethality means the heterozygotes are wild type.
2. Dominantly lethal alleles cannot be inherited.

Multiple Choice

3. Lethality that is associated with one or two genotypes occurs because
   a. the gene product is necessary for a vital process.
   b. the phenotype is a target for predators.
   c. the phenotype of the parent is sterile.
   d. the different phenotypic classes compete and one consistently loses.
4. A mating between two similar-phenotype organisms produces large numbers of two kinds of progeny in a 2:1 phenotypic ratio. The reason is
   a. a small sample size is causing random deviation from the 3:1 ratio.
   b. one allele is recessively lethal.
   c. one allele is dominantly lethal.
   d. one allele is incompletely dominant.
   e. one genotype is incompletely penetrant.

Short Answer

5. A series of crosses between a yellow mouse and a darker mouse produced 28 yellow and 31 dark mice. If you suspected one of these alleles to be recessively lethal, how would you determine which one? Be specific.

Topic Test 3: Answers

1. **False.** The heterozygotes may be wild type, as in cystic fibrosis, or they may have another phenotype, as in Manx cats.
2. **False.** These alleles are heritable when they cause death in adults.
3. **a.** Insufficient or malfunctioning gene product causes an essential process to fail in the case of recessive lethality. Interfering gene product wrecks an essential process in the case of dominant lethality.
4. **b.** Recessive lethality means that one of the homozygous classes from a monohybrid cross is dead and therefore missing from the phenotypic ratio. This converts a 1:2:1 ratio into a 2:1 ratio.
5. A good approach would be to cross the yellow mice to each other and the dark mice to each other. One of these crosses will show the parents to be true breeders (in fact, the dark ones). The other cross would give both phenotypes in a 1:2 progeny ratio (twice as many yellow as dark mice). The parents in this latter cross are heterozygous for the lethal allele.

TOPIC 4: MULTIPLE ALLELES

KEY POINTS

✓ *How many alleles are possible in the population?*

✓ *How do the alleles behave with respect to each other?*

✓ *How many alleles can each organism have per gene?*

Until now, we have discussed only two alleles per gene, for example, round and wrinkled alleles for pea seed shape. Most of the organisms used for formal genetic studies are **diploid**, as are humans, so each individual has only two alleles for any gene. But this does not mean that only two alleles exist in the population. Let us say that one individual is heterozygous for a gene. A different individual may have either or both of these two alleles, have a third allele and one of the previously known alleles, be homozygous for the third allele, have a fourth allele in combination with any of the other three, and so on. The possibilities are limited only by the total number of alleles in the population. Still, any one organism has only two alleles of any gene, one on each of the two homologs that carry that gene.

Let us return to the example of human blood groups, this time the ABO blood group system, whose name comes from the fact that there are three alleles for this gene, although there are only two in any one person. People can be type A, B, AB, or O phenotypically. The genotypes are a bit more complicated because of the special relationship between these alleles. Look carefully at the genotypes in **Table 3.1**. You should notice that I^A and I^B are codominant, and both are dominant to i. Again, any one person has only two alleles, but three exist for the ABO blood group gene. Note that no matter how many alleles there are, their behavior in meiosis is still consistent with Mendelian principles. For example, a mating of an AB-type person with an O-type person is genotypically $I^AI^B \times ii$, and produces A-type (I^Ai) and B-type (I^Bi) progeny in equal proportions. Multiple alleles can exist for any gene, and the two alleles in an individual can exhibit complete dominance, incomplete dominance, codominance (as in the ABO blood group case), or lethality.

Table 3.1 Phenotypes and Genotypes for Human Blood Groups

PHENOTYPE	GENOTYPE
A	I^AI^A or I^Ai
B	I^BI^B or I^Bi
AB	I^AI^B
O	ii

Topic Test 4: Multiple Alleles

True/False

1. However many alleles there are of a gene, a diploid organism can have only two.
2. Multiple alleles of a particular gene may exhibit all of the previously described Mendelian and non-Mendelian properties.

Multiple Choice

3. The fruit fly, *Drosophila melanogaster*, normally has brick-red eyes, but variants whose eyes are white, buff, apricot, or cherry have been observed. Crosses between these mutant flies produce a variety of eye colors (but never brick red). What could explain all of these related phenotypes?
   a. Monohybrids and dihybrids
   b. Incomplete dominance
   c. Multiple alleles
   d. Failure of random segregation
   e. Incomplete penetrance
4. What is the maximum number of different phenotypes that could be produced by the mating of a blood type AB individual to a type B individual?
   a. 1
   b. 2
   c. 3
   d. 4

Short Answer

5. There are four babies with blood types A, B, O, and AB waiting to be reunited with their parents. Can these children be correctly assigned to their parents solely on the basis of ABO blood typing if the pairs of parents are O and AB, A and B, A and A, and O and O? Explain.
6. Three lines of tulips are white, red, and purple. Crosses of red tulips with purple tulips result in black-flowered progeny. Crosses between the white and red lines produce red progeny. What can you tell about the flower-color character from these crosses?

Topic Test 4: Answers

1. **True.** That is what being diploid means. The phrase "multiple alleles" refers to the number of alleles in the population.
2. **True.** There can be complex dominance and codominance relationships between the alleles or they can behave as simply as Mendel's alleles.
3. **c.** These are four alleles of the same gene. They all affect the same set of traits (eye color) and show various dominance relationships in that they produce various eye colors, except the wild type.
4. **c.** There are two possible genotypes for the B person, I^Bi and I^BI^B. The first of these makes two kinds of gametes while the second makes only one. So the more diverse progeny come from the mating of $I^AI^B \times I^Bi$, with three phenotypic classes exhibiting the ratio 1 AB : 2 B : 1 A. (Note that this 1 : 2 : 1 ratio does not mean what 1 : 2 : 1 has meant so far in this text, i.e., the cross of monohybrid heterozygotes.)
5. Yes. The parents of each child are unambiguous. The AB child must belong to the A and B parents and the O child must belong to the O and O parents. The AB and O parents could have either the A or the B child, but the A and A parents cannot have the B child so must have the A child, leaving the B child to the AB and O parents.
6. The first cross suggests that red and purple alleles are codominant. The heterozygote makes both red and purple pigments, giving a darker flower. The second cross shows red to be dominant to white. Since the red flowers showed a dominance relationship with both the white and the purple flowers, they are all alleles of the same gene.

TOPIC 5: GENE INTERACTION

KEY POINTS

✓ *What phenotypic ratios arise from gene interaction? How is it recognized?*

✓ *Which Mendelian ratio is affected?*

Mendel's world was artificially simple: one gene, one character. However, it actually takes anywhere from one to scores of genes to make each aspect of life occur. You already know what ratios to expect from dihybrid crosses when the genes affect different processes, but what happens to the ratios when the two segregating genes in the dihybrid affect the *same* character? The short answer is that **interaction** between the genes will produce a modified phenotypic ratio. The type of modification indicates the sort of interaction between the genes. (The phrase *gene interaction* is the standard terminology but is also a misnomer. It isn't the genes but rather the gene products that interact, either directly or indirectly, so that both genes affect a single character.)

Our first example is **recessive epistasis**. Labrador retrievers have three coat colors: black, brown (which breeders call chocolate), and golden. Crosses between a particular pair of black dogs produce three progeny classes in the ratio 9 black : 3 brown : 4 golden. This ratio poses a bit of a puzzle: Is it a modified 9 : 3 : 3 : 1 ratio (i.e., 9 : 3 : 3+1) or is it a 1 : 2 : 1 ratio with small numbers of progeny causing deviation? For large progeny numbers, chi-square analysis will reveal a better fit with the 9 : 3 : 4 ratio (hypothesis of two genes showing recessive epistasis) than with the 1 : 2 : 1 ratio (hypothesis of one gene, two alleles showing incomplete dominance).

Table 3.2 Progeny of the Cross *MMDdWw* × *MMDdWw*

GENOTYPE	PHENOTYPE
9/16 *MMD-W-*	white with red spots
3/16 *MMddW-*	white with red spots
3/16 *MMD-ww*	entirely dark red
1/16 *MMddww*	entirely red

Since the 9 : 3 : 4 ratio is a modification of 9 : 3 : 3 : 1, let us hypothesize that the black parents are dihybrids (e.g., *BbEe*, where *B* and *E* are two genes that affect coat color) and the progeny are in the general classes 9 *B-E-* : 3 *bbE-* : 3 *B-ee* : 1 *bbee*. This genotypic ratio is modified into the 9 : 3 : 4 phenotypic ratio by the gene interaction that causes the animals that are *ee* genotypically to be the same golden phenotype, regardless of the *B* genotype. The last two genotypic classes combine into a single phenotypic class to give a 9 : 3 : 4 ratio, with *B-ee* and *bbee* together comprising the "golden" class. Dogs that are *B-E-* are black and those that are *bbE-* are brown. This is called **recessive epistasis** (*epistasis* means "suppression") because a recessive genotype for the *E* gene masks the phenotype of the *B* gene and causes a new, different phenotype (golden) in place of the *B* phenotypes. Molecularly, this can be explained by the *B* gene product being responsible for the production of a black pigment; in dogs that are *BB* or *Bb*, the pigment is made and in *bb* dogs no black is made so the dogs are brown. The *E* gene product causes the pigment to be deposited in the hair. If the dog is *B-E-*, it has black hair because it can make the pigment (*B-*) and can deposit it in the hair (*E-*). Dogs who are *bbE-* still make a brown pigment (but not the black one) and can deposit it in the hair, giving them brown hair. Dogs who are *ee* cannot deposit either black or brown into the hair, so the hair appears golden (its "natural" color). Since the hair is neither black nor brown, the *B* genotype is being masked in the hair by the *ee* genotype. We say that the *E* gene is epistatic to the *B* gene for hair color, because the genotype at *E* can mask the genotype at *B*. Interestingly, the genotype of the *E* gene only affects the hair—the skin of *ee* animals is black in *BB* and *Bb* animals but brown in *bb* animals. The *B* genotype is discernible in golden dogs from the color of the skin around their noses and lips because the pigment is made in the skin.

Dominant epistasis is another form of gene interaction, demonstrated in the flower *Digitalis purpurea* (foxgloves). Wild-type flowers are white with red spots in the throat of the flower. Three genes are relevant to our discussion of this phenotype. The *M* gene determines the ability to make the red pigment. *M-* plants can make red pigment, while *mm* plants cannot and so their flowers are completely white. The *D* gene product is an enhancer of the pigment; *D-* plants have darker red flowers than *dd* plants (assuming both are *M-*). Finally, the *W* gene determines the location of the pigment. The wild-type gene product restricts the pigment to spots in the flower's throat so that *W-* flowers are white with red spots but *ww* flowers are completely red. We will look at a simple cross involving dihybrids of the genotype *MMDdWw*. The flowers of these plants are white with red spots, a phenotype that can be explained by the *M* gene allowing pigment to be produced and the *W* allele restricting the pigment to the spots. The progeny of this dihybrid cross are shown in **Table 3.2**. Notice that the *MMD-W-* and *MMddW-* progeny have the same phenotype, so the apparent ratio is actually 12 : 3 : 1. This is the dominant epistasis ratio, indicating the ability of a dominant genotype (in this case, *W-*) to mask the phenotype from another locus (such as the *D-* or *dd* genotype) and replace it with a different phenotype (color in spots). Incidentally, the effect of *mm* on *D*- is also recessive epistasis, demonstrating that more than two genes can show epistasis and that the interactions need not be of a single type.

Other kinds of gene interaction produce other ratios, such as 9:7, indicating duplicate gene function, or 9:3:3:1 (four distinct phenotypes but *only one* character is affected, as for chicken combs of single, walnut, pea, and rose shape). All result in modification of the classic dihybrid F_2 ratio 9:3:3:1 because two genes are involved in producing the variety of phenotypes seen for a single character. The common characteristic of all of these examples is that the pairs of genes being considered work together to determine the phenotype of a single character. As before, Mendel's principles of segregation and independent assortment are still observed.

Topic Test 5: Gene Interaction

True/False

1. The masking of one gene's phenotype by another gene is called epigenetics.
2. Gene interaction is demonstrated by modification of the 9:3:3:1 ratio seen in dihybrid crosses.
3. There are only two ways in which the dihybrid Mendelian ratio can be modified and still be gene interaction.

Multiple Choice

4. The progeny of the dihybrid cross *AaBb* × *AaBb* exhibit three phenotypes in the ratio 9:3:4. What does this tell you?
   a. One of the genotypes is lethal.
   b. Too few progeny have been analyzed to see the 1:2:1 ratio.
   c. The heterozygous class is showing incomplete penetrance.
   d. The genes did not undergo independent assortment.
   e. One of these genes is recessively epistatic to the other gene.
5. Two chickens with walnut-shaped combs were crossed to yield a surprising bunch of progeny. There were four phenotypic classes in the proportions 9/16 walnut, 3/16 pea, 3/16 rose, and 1/16 single. What genetic explanation accounts for chicken-comb inheritance?
   a. Dominant epistasis
   b. Recessive epistasis
   c. One gene, four phenotypes
   d. Two genes affecting single character
   e. Two genes affecting two characters

Short Answer

6. A white-fruiting summer squash was allowed to self. The seeds it produced developed into 59 white-fruiting plants, 19 yellow-fruiting plants, and 5 green-fruiting plants. What caused these results? Assign genotypes to the phenotypic classes.

7. A black mouse from a true-breeding line was mated to an albino mouse, also from a true-breeding line, and all of the progeny were agouti (hairs have alternating bands of black and yellow, giving the animal an overall gray appearance). Crosses of the agouti F_1 animals to each other result in offspring that show all three phenotypes in the proportions 9/16 agouti, 3/16 black, and 4/16 albino. What genetic phenomenon are these crosses demonstrating? Give genotypes for all individuals.

Topic Test 5: Answers

1. **False.** The correct word is *epistasis.*
2. **True.** Interaction can be seen when two genes are segregating and this requires dihybrid crosses, at least, so the ratio that will be modified is 9:3:3:1. Gene interaction can also be seen among three or more genes.
3. **False.** There are as many gene interactions as there are modifications of the 9:3:3:1 ratio. For completely dominant alleles of both genes, phenotypic ratios of 9:3:4, 12:3:1, 9:7, 9:6:1, 15:1, 13:3, and 9:3:3:1 can be seen.
4. **e.** The compression of the 1/16 *aabb* class into one of the 3/16 classes is the result of recessive epistasis, either of *b* on the *A* gene or of *a* on the *B* gene.
5. **d.** A single character is being affected and the ratio clearly implies two genes segregating. This form of gene interaction involves dominant alleles of two genes producing a new phenotype: walnut.
6. These progeny are exhibiting a modified 9:3:3:1 dihybrid ratio, 12:3:1. There must be interaction between these two genes such that one of the 3/16 classes looks like the 9/16 class. Let *C* stand for the wild-type (colorless) allele of the *color* gene and *c* for the mutant (color) allele. *G* will be the wild-type (not green) allele of the *green* gene; *g* is the mutant allele. The squashes in the 12/16 class are white and are *G-C-* or *ggC-*, those in the 3/16 class are yellow and *G-cc*, and those in the 1/16 class are green and *ggcc*. So *C* is epistatic to the *G* gene. Perhaps the *C* gene product normally inhibits the other colors to keep the fruits white, and in its absence yellow or green is produced, where they normally would not be.
7. The parents in the cross are homozygotes and since they produce an animal that breeds like a dihybrid, they must be homozygous recessive for different genes. The black animal is *aaCC* and the albino is *AAcc*, where *A* and *a* are alleles of the agouti gene and *C* and *c* are alleles of the colorless gene. Their progeny are *AaCc* and are agouti, the wild-type coat color for many mammals. These dihybrids have progeny that are 9/16 *A-C-* (agouti), 3/16 *aaC-* (black), and 4/16 *A-cc* or *aacc* (albino). The recessive *cc* genotype is epistatic to the *A* gene.

TOPIC 6: SEX LINKAGE

KEY POINTS

✓ *What is sex linkage?*

✓ *What experiments proved the chromosomal theory of inheritance?*

✓ *Was this also proof of sex linkage?*

This topic explores the violation of Mendel's principle of independent assortment by a phenomenon known as **linkage**. We will study linkage in greater detail in Chapter 4 but will cover its discovery here, because a particular type of linkage, called **sex linkage**, proved the **chromosomal theory of inheritance**. The story begins with Thomas Hunt Morgan, whose laboratory at Columbia University (1904–28) was the notorious "Fly Room." Morgan had many bright people working for him, including H. J. Muller, Calvin Bridges, and Alfred Sturtevant, who performed the experiments that will be described here and in Chapter 4. Morgan was interested in development in the fruit fly *Drosophila melanogaster* when, out of curiosity, he began to work with a white-eyed fly that had appeared in his stocks.

His observations were among the first of the phenomenon of linkage, and he was the first to explain it. He recognized the behavior of this mutation to be a consequence of sexually dimorphic chromosomes and thereby verified the chromosomal theory of inheritance, which had previously been supported only by circumstantial evidence. Wild-type flies of this species have brick-red eyes. The fly that Morgan noticed in his stocks had completely white eyes and, it is important to note, was male. Morgan crossed this fly to a wild-type (red-eyed) female and collected F_1 flies. All of the F_1 offspring were red-eyed, indicating that white is recessive. He allowed the F_1 progeny to mate with each other and analyzed the F_2 generation. Overall, this generation exhibited the expected 3 red : 1 white ratio; however, all of the white-eyed flies were male! The red-eyed flies in this generation were in the ratio 2 females : 1 male.

It was quite a surprise that color and sex seemed to be related in their inheritance. A testcross of the F_1 red-eyed females to white-eyed males resulted in progeny in the ratio 1 red-eyed female : 1 red-eyed male : 1 white-eyed female : 1 white-eyed male, confirming that the F_1 females were heterozygous. Morgan also performed **reciprocal crosses** (i.e., switched the sexes of red-eyed and white-eyed flies), to see whether it mattered which parent passed on the recessive trait. Morgan mated a white-eyed female to a red-eyed (true-breeding) male, and observed the ratio 1 red-eyed female : 1 white-eyed male. This trait did not behave the same in a reciprocal cross!

Reciprocity is true of autosomal traits; since this trait is not reciprocal, Morgan reasoned, it must be unequally represented in the male and female flies. He knew that *Drosophila* have **sexually dimorphic chromosomes** (i.e., chromosomes that differ between the sexes): Females have two X chromosomes while males have both an X and a smaller Y chromosome. (The sex with dissimilar sex chromosomes is the **heterogametic sex**.) Morgan realized that these sexually dimorphic chromosomes determine sex. Furthermore, he concluded that the X chromosome in particular must carry genes, such as the white gene, that would be unequally represented in the two sexes. This last bit of reasoning was based on the visual dissimilarity between the X and Y chromosomes and the fact that it explained the special inheritance pattern of the white gene. (Aside: Geneticists often refer to genes, not by the character they influence, but by the mutant [or recessive] phenotype. In this case, the gene in question is usually referred to as the "white gene.")

Let X^w represent the white allele and X^W the red allele; lowercase is used for the recessive trait as usual but is used as a superscript on an X chromosome symbol (for now) to reinforce the knowledge that the alleles reside on the X chromosome. Using the X and Y symbols should help you to avoid mistakes that come from confusing it with an autosomal trait. The original cross was X^WX^W females with the X^wY male. Their progeny can be observed in **Figure 3.2**. Diagram the reciprocal cross for yourself (red-eyed male × white-eyed female), to see that the creation of different progeny in reciprocal crosses is the flashing red light for sex-linked traits.

One last comment about sex-linked traits should be made: There are a number of traits in mammals that may appear to be sex-linked but are not, like milk production in mammals, horns

	X^w	Y
X^W	X^WX^w red female	X^WY red male
X^W	X^WX^w red female	X^WY red male

Figure 3.2 Punnett square for a cross between X^WX^W females and X^wY males.

in sheep, or bald spots in humans. **Sex-limited traits**, like milk production, are traits in which the expression of the phenotype is limited to one sex, for example, by hormones. No matter what sort of a cross is tried, the nonexpressing sex will continue not to express the trait. **Sex-influenced traits**, like horns and hair loss, are traits where the sex of the individual influences the expression of the phenotype but that phenotype is not limited to one sex. For these traits, expect to see the heterozygotes having different phenotypes depending on their sex, but the homozygotes will have the same phenotype regardless of sex. Take care not to be confused by these autosomal traits.

Topic Test 6: Sex Linkage

True/False

1. Sex-linked traits show non-Mendelian inheritance as a result of different numbers of alleles present in males versus females.
2. A male who has a dominant X-linked disorder will have affected sons and unaffected daughters.

Multiple Choice

3. If the white-eyed fly Morgan noticed in his stocks had been female, in what generation would he have noticed its deviation from the Mendelian expectation?
   a. In the F_1 generation when the males and females had different eye colors.
   b. In the F_2 generation when the males and females had different eye colors.
   c. In the F_1 generation when the males and females had the same eye color.
   d. In the F_2 generation when the ratios were unexpected.
   e. More than one of the above answers is correct.
4. Morgan's great discovery that was described in this topic was that
   a. fly traits do not behave as expected in reciprocal crosses.
   b. flies could have white eyes.
   c. flies determine sex by their eye color.
   d. flies have genes on their X chromosomes.
5. In cattle, males are XY and females are XX. If a female cow is heterozygous for a recessive sex-linked trait and she is mated to a wild-type male, what progeny should be expected?

a. Normal females, and normal and affected males
b. Normal females and affected males
c. Normal and affected males and females
d. All normal progeny
e. All affected progeny

Short Answer

6. Use a Punnett square to show the reciprocal cross that was suggested in the topic text. How do the progeny differ from those in the original cross?

Topic Test 6: Answers

1. **True.** For example, human females have two alleles of every X-linked gene, while human males have only one. (Males are said to be **hemizygous**.)
2. **False.** The man will have affected daughters and normal sons, because the sons get their only X chromosome from their mothers.
3. **e.** Answers a and d are both correct. The F_1 males and females had different eye colors, and the F_2 males and females had both eye colors, but the ratio was surprisingly like that of a dihybrid testcross.
4. **d.** This topic introduces sex linkage as it was discovered, at which time the presence of genes on the sex chromosomes was proved. The proof was that the behavior of the white gene exactly paralleled the behavior of the sexually dimorphic chromosomes.
5. **a.** $X^AX^a \times X^AY$ gives 1/2 normal females, 1/4 normal males, and 1/4 affected males.
6.

	X^W	Y
X^w	X^WX^w red female	X^wY white male
X^w	X^WX^w red female	X^wY white male

The results of this cross (white-eyed female to red-eyed male) are 1/2 red-eyed females and 1/2 white-eyed males. Morgan's original cross (white-eyed × red-eyed female) gave 1/2 red-eyed females and 1/2 red-eyed males. The phenotype of the male progeny is different in the two crosses.

IN THE CLINIC

There are numerous examples in medicine of non-Mendelian genetic phenomena like the ones described in this chapter. Many of us are familiar with the painful, debilitating disease sickle-cell anemia which disproportionately affects people of African (and some of

Indian) descent. This disease is an example of recessively lethal alleles that do not give a wild-type phenotype in the heterozygote; one wild-type allele is sufficient for survival, but two wild-type alleles are required for a wild-type phenotype. The recessive sickle-cell allele specifies an alternative form of hemoglobin, the molecule that carries oxygen in the bloodstream. Homozygous recessive individuals have a severe anemia (dearth of red blood cells owing to their reduced life spans) that causes reduced quality of life, reduced fertility, and premature death. Heterozygotes are mostly normal, except under conditions of low oxygen, when anemia occurs. Thus, they are not completely wild type in this, and in another way: Heterozygotes are resistant to infection by the parasite *Plasmodium falciparum* which causes malaria, while wild-type homozygotes are not resistant. This selection is the basis for the prevalence of the recessive allele in places where malaria is endemic. Thus, sickle-cell anemia is an example of a recessively lethal allele that also has a nonlethal dominant phenotype.

DEMONSTRATION PROBLEM

Question: In a mysterious tropical fish there are three phenotypes associated with a single character. The alleles *S* and *s* cause puffy and spiny fish, respectively. The *G* allele of a second, unlinked gene causes the entire fish to be enormous, regardless of the genotype of the *S* gene. The cross of an enormous fish to a spiny fish produced 1/2 enormous, 1/4 puffy, and 1/4 spiny progeny. How do you account for the phenotypes of parents and results of this cross?

Answer: Dissect this problem one sentence at a time. "Three phenotypes" leads one to expect a modified Mendelian ratio, perhaps 1:2:1 or one of the epistasis ratios. The next sentence describes a gene with two alleles that are apparently simple: *S* specifies the puffy phenotype (genotypes *SS* and *Ss*) and *s* specifies spiny (genotype *ss*). The third sentence introduces a second gene, *G*; therefore, one of the epistasis ratios should be expected. This new gene causes a third phenotype (supporting the epistasis conclusion) when the dominant allele, *G*, is present; this must be dominant epistasis. The ratio to expect from a dihybrid cross is 12 *G---*: 3 *ggS-*: 1 *ggss*, where the phenotypes are enormous, puffy, and spiny, respectively. The cross was enormous × spiny and now we know that spiny is doubly recessive, *ggss*. To get any nonenormous progeny, the enormous parent must be heterozygous for the *G* locus (*Gg*), so that half of the progeny will not be enormous. To get 1/4 each of puffy and spiny progeny, the enormous parent must have one dominant *S* allele and one recessive *s* allele. Therefore, the cross was *GgSs* × *ggss*.

Chapter Test

True/False

1. Dihybrid ratios of 1:2:1 where 1/2 of the progeny have intermediate phenotypes suggest incomplete dominance of alleles.
2. Gene interaction can be recognized by modification of the monohybrid 3:1 ratio.
3. Lethal alleles can only be inherited if the lethality is recessive.
4. Pedigrees are used to collect information about fortuitous matings that exhibit interesting traits.

5. The term *multiple alleles* refers to the genetic complexity of the population.
6. The term *gene interaction* refers to genes interacting with the environment to produce new phenotypes.
7. The lethal phenotype of a lethal allele affects embryos.
8. A mating between two people who are blood type AB creates offspring who have four blood type alleles each.

Multiple Choice

9. If a pedigree appears to show the inheritance of an autosomal dominant trait,
   a. affected offspring will have unaffected parents.
   b. mostly males will be affected.
   c. the mating of an affected person with an unaffected person produces 3/4 affected and 1/4 normal offspring.
   d. the trait will appear to "skip" generations.
   e. most affected people will have some affected children.
10. Two black mice were mated to produce 10 brown, 13 gray, and 29 black offspring. Some of the black offspring were true breeders. What could explain these results?
    a. Dominant epistasis
    b. Recessive epistasis
    c. Incomplete dominance
    d. Lethal alleles
    e. Independent assortment
11. Which statement is true?
    a. The phrase *multiple alleles* means each individual has more than two alleles.
    b. Multiple alleles result from matings between dissimilar true breeders.
    c. Each of the multiple alleles of a gene can have simple or complex interactions with other alleles.
    d. All of the above are true.
    e. None of the above are true.
12. A man has Fabry disease, a recessive X-linked disorder causing kidney, eye, and skin problems. His daughter is normal and she wants to know the likelihood she will have a child with Fabry disease. Her husband is normal. What is the likelihood?
    a. 2/3 for each child
    b. 1/2 for each child
    c. 1/4 for each child
    d. 1/8 for each child
    e. 0 for each child
13. A male and female *Drosophila* have kidney-shaped eyes consisting of about 350 facets each. When they are mated, they produce three kinds of progeny: 121 have round eyes of about 800 facets each, 140 have bar-shaped eyes of about 50 facets each, and 254 have kidney-shaped eyes. What is the likely mode of inheritance for this trait?
    a. Incomplete dominance
    b. Recessive epistasis

c. Dominant epistasis
d. Simple dominance and recessiveness
e. Multiple alleles

14. When two individuals are homozygous for a disease allele but one is severely affected and one is mildly affected, we say the disease exhibits
a. variable expressivity.
b. incomplete penetrance.
c. incomplete dominance.
d. gene interaction.
e. recessiveness.

15. In pedigree analysis, an apparently dominant trait will occasionally be seen to skip a generation. This can be evidence for
a. recessiveness.
b. incomplete penetrance.
c. incomplete dominance.
d. lethal alleles.
e. X-linked inheritance.

Short Answer

16. Fragile X syndrome is the most common form of inherited mental retardation, yet individuals who have this disease can be mildly retarded, can be severely retarded, or can have any degree of retardation in between. What genetic phenomenon explains this observation?

17. What is the most likely mode of inheritance in the following pedigree? Assume the trait is rare and give evidence for your answer.

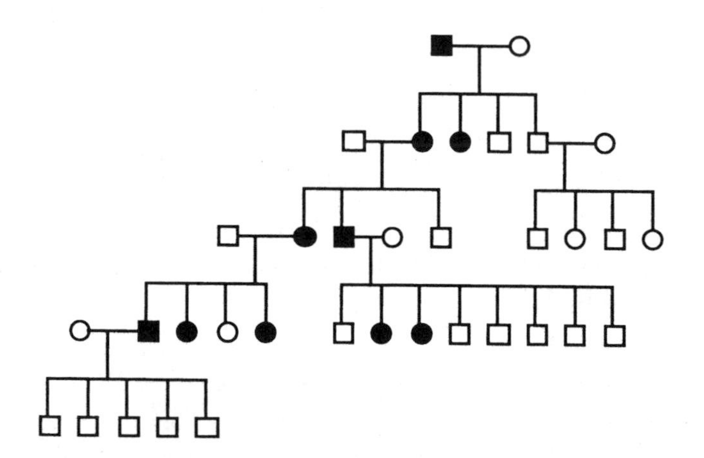

18. A true-breeding white horse is repeatedly mated to a true-breeding brown horse and all of the progeny are golden (palomino). What are two possible causes of the different phenotypes and how would you determine which one is correct?

19. How can a lethal allele be dominantly lethal?

20. Describe at least two crosses you could perform to test whether a newly identified mutation is sex-linked.

Essay

21. For each of the modes of inheritance described in this chapter, explain in what way each is inconsistent with Mendel's observations.

22. Summarize the new ratios from this chapter and what each one means.

Chapter Test Answers

1. **T** 2. **F** 3. **F** 4. **T** 5. **T** 6. **F** 7. **F** 8. **F** 9. **e** 10. **b** 11. **c** 12. **c** 13. **a** 14. **a** 15. **b**

16. Variable expressivity.

17. This pedigree is most likely tracking a dominant X-linked trait. The best evidence for this hypothesis are the affected males in generations III and IV who have affected daughters but normal sons.

18. Incomplete dominance is the real reason for golden horses. This can be tested by crossing the golden horses with each other. Their progeny will be 1/4 white, 1/2 golden, and 1/4 brown. Another possibility for all golden progeny could be epistasis. Suppose white (homozygous *ww*) is recessively epistatic to color (*W-*) and brown (homozygous *bbW-*) is recessive to golden (*B-W-*); this mating could have been *BBww* (white) × *bbWW* (brown) to produce *BbWw* (golden). Crossing the golden horses together would result in 9/16 golden, 3/16 brown, and 4/16 white offspring.

19. Dominantly lethal alleles can be inherited if the lethal phenotype occurs after the onset of childbearing. If the lethality is embryonic or in childhood, dominant lethal alleles only exist as new mutations.

20. Sex-linked traits behave differently in reciprocal crosses. Cross a female with the trait to a wild-type male, and cross a wild-type female to a male with the trait. If there are similar progeny from the two crosses, the trait is not sex-linked. Alternatively, the F_1 offspring from cross of a wild-type female to a mutant male could be crossed to each other. If the trait is sex-linked, the F_2 generation will reveal a sex-specific segregation pattern.

21. Incomplete dominance and codominance extend the dominance and recessiveness concepts and replace Mendel's monohybrid phenotypic ratio with the genotypic ratio, 1:2:1. Lethal alleles do not directly violate a Mendelian principle but appear to do so by producing a modified F_2 ratio, 2:1, when the homozygous lethality occurs before birth. Gene interaction modifies Mendel's dihybrid ratio, but is consistent with the independent assortment principle. It actually complicates the dominance and recessiveness concepts. Sex-linkage violates Mendel's independent assortment principle.

22. Modified 3:1 ratios represent incomplete dominance or codominance if they are 1:2:1 or a lethal allele if they are 2:1. In either case, the most numerous class is the heterozygote and the parents of this generation were monohybrids. Modified 9:3:3:1 ratios indicate gene interaction in the progeny of dihybrid parents. The 9:3:4 ratio indicates recessive epistasis, where the 4/16 class has the masking phenotype. The 12:3:1 ratio indicates dominant epistasis, where the 12/16 class has the masking phenotype.

Check Your Performance:

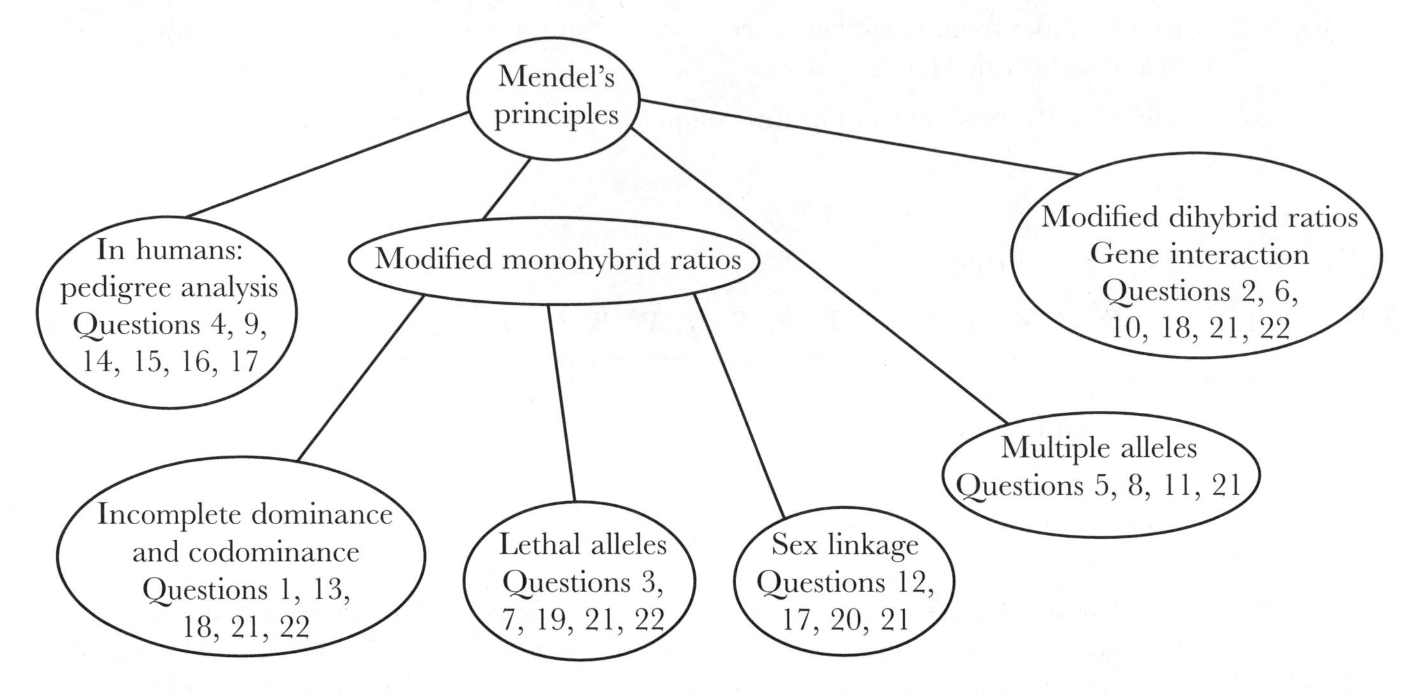

Use this chart to identify weak areas, based on the questions you answered incorrectly in the Chapter Test.

First Midterm Exam

True/False

1. The monohybrid cross does not provide evidence of independent assortment.
2. If 1 out of every 10,000 babies born in a particular city has the disease phenylketonuria, and if one affected baby were just born, then the next baby born will definitely not have the disease.
3. Sex-limited traits show non-Mendelian inheritance as a result of different numbers of alleles present in males versus females.
4. Recessive lethality means the heterozygotes are dead.
5. Human families usually show Mendelian ratios.
6. The progeny from a cross of dissimilar true breeders are heterozygotes.
7. The product rule is useful for calculating the probability of two independent events occurring at the same time.
8. Different behavior of a trait in reciprocal crosses is evidence of X-linkage.
9. Meiosis is not important to Mendel's principles.
10. In a cross of dissimilar homozygotes, the first indication that a trait is incompletely dominant is the F_2 phenotypic ratio.
11. Progeny ratios from many crosses can be used to infer the genotypes of the parents from each cross.

Multiple Choice

12. Which is not true of incomplete dominance?
    a. The phenotype of the heterozygote is intermediate between the phenotypes of the homozygotes.
    b. It can be recognized by a 1:2:1 F_2 ratio.
    c. It reflects intermediate levels of active gene product.
    d. It is symbolized with capital letters for all incompletely dominant alleles.
    e. It occurs when the environment is conducive.
13. What is wrong with the following pedigree?

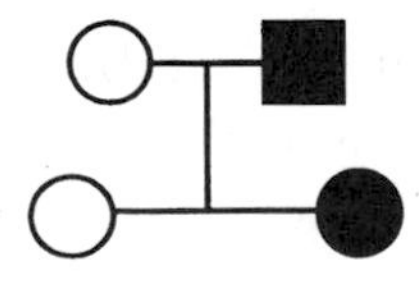

    a. Wrong symbols for male and female
    b. Too few people
    c. Generation I mating displayed incorrectly
    d. Generation II progeny displayed incorrectly
    e. None of the above

14. Which of the following is not true of testcrosses?
    a. Progeny that comprise two equivalently sized classes indicate one parent is heterozygous.
    b. Progeny that have the same phenotype indicate that the "unknown" parent is homozygous.
    c. The progeny indicate dominance and recessiveness.
    d. One parent is homozygous recessive.

15. Which of the following is *untrue*? The chi-square test
    a. is for determining whether real data are consistent with a hypothesis.
    b. will recommend against a hypothesis, rather than disprove it.
    c. gives a probability that similar deviation from the expected ratios will occur if we did the same experimental crosses again.
    d. will tell us whether the real data represent a too small sample rather than not fitting the hypothesis.
    e. can be used repeatedly on the same set of data, testing different hypotheses.

16. A normal human cell has 46 chromosomes. How many do human gametes have?
    a. 11
    b. 23
    c. 46
    d. 92

17. Crosses between yellow snapdragons give 52 yellow and 23 green. Crosses of yellow snapdragons to green gives 32 yellow and 29 green. What could explain these results?
    a. Simple dominance of yellow alleles
    b. Incomplete dominance of yellow and green alleles
    c. Gene interaction resulting in yellow and green phenotypes
    d. Recessive lethality of yellow allele
    e. Codominance of yellow and green alleles

18. The cross of a dihybrid with a homozygous recessive results in a phenotypic ratio of
    a. 1:1.
    b. 3:1.
    c. 1:1:1:1.
    d. 9:3:3:1.
    e. 3:1:3:1.

Short Answer

19. In *Drosophila*, white eyes are caused by an X-linked recessive allele and curved wings are caused by an autosomal recessive allele. A female with curved wings is mated to a male with white eyes. Predict the phenotypes and ratios of the F_1 and the F_2 generations.

20. Two black pigs were mated. There were 10 progeny. Eight were black and 2 were pink. What can you conclude?

21. Red-green color blindness is an X-linked recessive trait in humans. If the father of a woman with normal vision was color blind, as is her husband, what is the probability that her firstborn child will be color blind?

22. A dominant allele causes people to have hairy palms and soles. If two heterozygotes for this allele had six children, how would you calculate the probability that at least three will be affected?

Essay

23. The selfing of a yellow round pea resulted in the following progeny: 53 yellow round, 21 yellow wrinkled, 19 green round, and 7 green wrinkled. Are these progeny in agreement with Mendel's dihybrid ratio? Test by chi-square analysis.

24. Distinguish between recessive epistasis and dominant epistasis.

25. Draw a pedigree for a family consisting of two sets of grandparents in generation I, each producing one male and one female child (in that order). The parents of the propositus are one of the second-generation females and the male from the other mating. The propositus is an albino male and his maternal grandmother is also albino. His paternal grandfather is deceased.

Answers

1. **T** 2. **F** 3. **F** 4. **F** 5. **F** 6. **T** 7. **T** 8. **T** 9. **F** 10. **F** 11. **T** 12. **e**

13. **d** 14. **c** 15. **d** 16. **b** 17. **d** 18. **c**

19. The mating was $ccX^WX^W \times c^+c^+X^wY$. F_1 progeny are 1/2 wild-type females ($c^+cX^WX^w$) and 1/2 wild-type males (c^+cX^WY). F_2 offspring are 6/16 wild-type females, 2/16 curved females, 3/16 wild-type males, 3/16 white males, 1/16 curved males, and 1/16 white curved males.

20. Black is dominant, pink is recessive, and the parents were both monohybrids.

21. For this answer, X^C is the wild-type allele of the color blindness gene, and X^c is the color blind allele. The woman must be heterozygous (X^CX^c) since she got her father's X^c chromosome. The chance that she will pass it on to her children is 1/2. Her husband is X^cY, so the chance of having an affected son is p(mother's X^c) × p(father's Y) = 1/2 × 1/2 = 1/4. The chance of having an affected daughter is p(mother's X^c) × p(father's X^c) = 1/2 × 1/2 = 1/4. Finally, the probability that the firstborn will be either an affected son or an affected daughter is 1/4 + 1/4 = 1/2.

22. The binomial approach should be used for each acceptable number of affected children. Calculate the probability for 3 affected, 4 affected, 5 affected, and 6 affected children separately, then add these probabilities together. An example of the single calculation for 4 affected children is $6!/4!2!\ (3/4)^4\ (1/4)^2 = 0.297$. The total probability is 0.963.

23. The genetic hypothesis is that all of Mendel's principles are obeyed, dominance is complete, and the parent was a dihybrid. Expected values are 56.25 yellow round, 18.75 yellow wrinkled, 18.75 green round, 6.25 green wrinkled. $\chi^2 = 0.55$, df = 3, $p > 0.9$; therefore, the null hypothesis should be accepted.

24. In recessive epistasis the phenotype specified by one gene is replaced by a new phenotype whenever a second gene is homozygous recessive. The characteristic ratio for a dihybrid cross is 9:3:4, where 1/4 of the progeny have the masking phenotype. Dominant epistasis

is recognized by the 12:3:1 dihybrid ratio. In this case, the masking phenotype occurs when one gene has a dominant genotype.

25. 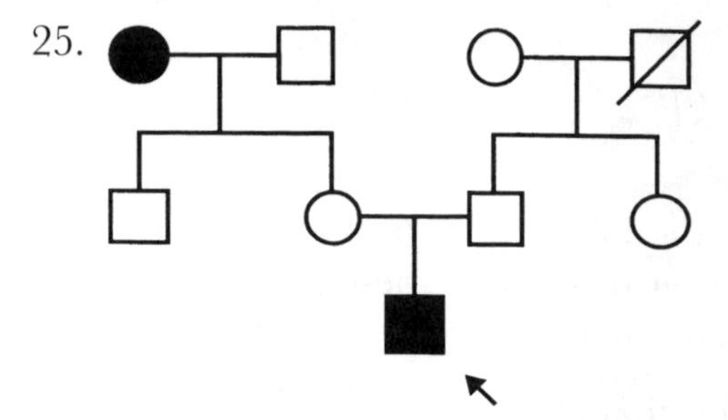

CHAPTER 4 Linkage

Sutton and Boveri postulated, and T. H. Morgan proved, that genes reside on chromosomes. It is estimated that a human cell contains 50,000 to 100,000 genes, but there are only 23 pairs of chromosomes in these cells. The yeast *Saccharomyces cerevisiae* is known to have about 6,000 genes on 16 chromosomes. Obviously there must be a large number, from hundreds to tens of thousands, of genes on each chromosome in all organisms. Yet Mendel saw genes exhibiting independent assortment. Either these genes must have been on different chromosomes, or there is a way for genes that are physically connected on a chromosome to become unconnected. In fact, both possibilities are true. Genes on separate chromosomes exhibit independent assortment. Genes on the same chromosome, said to be **linked**, will not exhibit independent assortment.

ESSENTIAL BACKGROUND

- **Independent assortment (Chapter 1)**
- **Testcrosses (Chapter 1)**
- **Gamete formation/meiosis (Chapter 1)**
- **Product rule (Chapter 2)**

TOPIC 1: BACKGROUND FOR LINKAGE

KEY POINTS

✓ *How is linkage detected?*

✓ *What are recombinant and parental types?*

The hallmark of independent assortment is the dihybrid selfing 9:3:3:1 and testcross 1:1:1:1 ratios (review Chapter 1 if this is not familiar). The first observation of a violation of this principle occurred in 1905 when William Bateson, E. R. Saunders, and R. C. Punnett selfed sweet peas that were heterozygous for a flower-color gene and for a pollen-shape gene (i.e., they were dihybrids) and did not observe the expected 9:3:3:1 F_2 ratio. Nor were their data consistent with any modification of this ratio by gene interaction. They could not explain what had happened to produce the new ratio, probably because they had selfed the dihybrid, which resulted in progeny that were too genotypically complex to sort out. If they had known what to look for, they could have recognized the linkage in their ratio. Several years later, T. H. Morgan performed a similar experiment, but instead of selfing the dihybrid, he testcrossed it. The beauty of the testcross is that for each phenotype in the offspring, there is only one genotype. Therefore,

	0.25 *AB*	**0.25 *ab***	**0.25 *Ab***	**0.25 *aB***
1.0 *ab*	0.25 *AaBb*	0.25 *aabb*	0.25 *Aabb*	0.25 *aaBb*

Figure 4.1 Punnett square depicting the expected frequencies from a testcross of *AaBb*.

this modification of the experimental approach was pivotal to discovering the cause of Bateson, Saunders, and Punnett's modified F_2 ratio: failure of the genes to undergo independent assortment. To see why Morgan succeeded with the testcross, consider **Figure 4.1**, a Punnett square, modified to include frequencies, that depicts the testcross of *AaBb* with *aabb*. A single row is used to represent the doubly recessive parent because it only makes one kind of gamete. This is also why the frequency of this gamete is 1.0. There are four kinds of gametes from the dihybrid. Since the genes undergo independent assortment, the frequency of each kind is 0.25. The frequencies of the progeny come from the product rule: The probability of each offspring being *AaBb* is the product of the probabilities of the *AB* gamete (0.25) and the *ab* gamete (1.0), which is 0.25. This is an important concept to master now, because when the genes are linked, the four kinds of gametes will not have the same frequencies. Another important concept is apparent if you imagine that the dihybrid in this example came from the cross *AABB* × *aabb*. Thus, the *A* and *B* alleles were in the same gamete and the *a* and *b* alleles were in the other gamete. This is the basis for classifying the dihybrid's progeny. If the dihybrid receives the *A* and *B* alleles together, then when the dihybrid passes them on to the next generation together, the resulting progeny are termed **parental**. There are always two parental-type gametes. For the dihybrid in this example, the parental gametes are *AB* and *ab*, and because this is a testcross, these are also the phenotypes of the parental-type progeny. The other two gametes *this* dihybrid produces, *aB* and *Ab*, are termed **recombinant**. These are gametes (and the resulting progeny) whose alleles are not in the same configuration as they were received by the dihybrid. For independent assortment to be true, the proportion of parental-type gametes must equal the proportion of recombinant-type gametes. This is easiest to determine by studying the progeny of a testcross.

Topic Test 1: Background for Linkage

True/False

1. Linkage is detected by deviation from the expected 1 : 1 : 1 : 1 progeny ratio in a testcross.
2. A dihybrid makes both parental and recombinant gametes.

Multiple Choice

3. A dihybrid is created by mating *AAbb* with *aaBB*. What parental-type gametes does this dihybrid make?
   a. *AaBb*
   b. *AB* and *ab*
   c. *Ab* and *aB*
   d. *Aa* and *Bb*
   e. *AB* and *ab* and *Ab* and *aB*

4. What recombinant-type gametes does the dihybrid in question 4 make?
   a. *AaBb*
   b. *AB* and *ab*
   c. *Ab* and *aB*
   d. *Aa* and *Bb*
   e. *AB* and *ab* and *Ab* and *aB*

Short Answer

5. If the cross in question 4 were *AABB* × *aabb*, and the dihybrid's gametes were analyzed, which would be parental, which would be recombinant, and how would you know whether the *A* and *B* genes are linked?

Topic Test 1: Answers

1. **True.** Independent assortment causes this ratio. Linkage is the absence of independent assortment.
2. **True.** And the ratios of all four gametes (and therefore both types) are equal if the genes are unlinked, and are unequal if the genes are linked.
3. **c.** The *A* and *b* alleles came in one gamete from one parent to the dihybrid, so *Ab* is parental; likewise the *a* and *B* alleles came to the dihybrid in one gamete, so *aB* is the other parental gamete.
4. **b.** The *AB* combination is new for a gamete in this family, as is the *ab* combination.
5. This dihybrid is *AaBb* but has the opposite set of parental and recombinant gametes to the one considered in questions 3 and 4. *AB* and *ab* are parental; *Ab* and *aB* are recombinant. If gene *A* and gene *B* are not linked, they will show independent assortment and a testcross will yield progeny in the ratio 1:1:1:1 and dihybrid selfings will produce progeny in the ratio 9:3:3:1. If these ratios are not obtained, the genes are linked.

TOPIC 2: LINKAGE AND CROSSING OVER

Key Points

✓ *Where do recombinant gametes come from?*

✓ *How do we describe allele configurations?*

✓ *What are loci? markers? crossover? recombination?*

Recombinant gametes are produced either by independent assortment if the segregating genes are physically contained on separate chromosomes, or by recombination (specifically, **crossover**) between homologs if the genes reside on the same chromosome. For the purpose of this description of meiotic recombination, assume the organism is diploid, and refer to **Figure 4.2**. In the S phase that precedes meiosis, each homolog is completely duplicated. After the S phase, the identical copies are known as **sister chromatids**. In early prophase I, one of these sister chromatids will be deliberately broken, and the broken ends will be used to synapse with the other homolog (also consisting of two sister chromatids), which will be used to repair the break. This

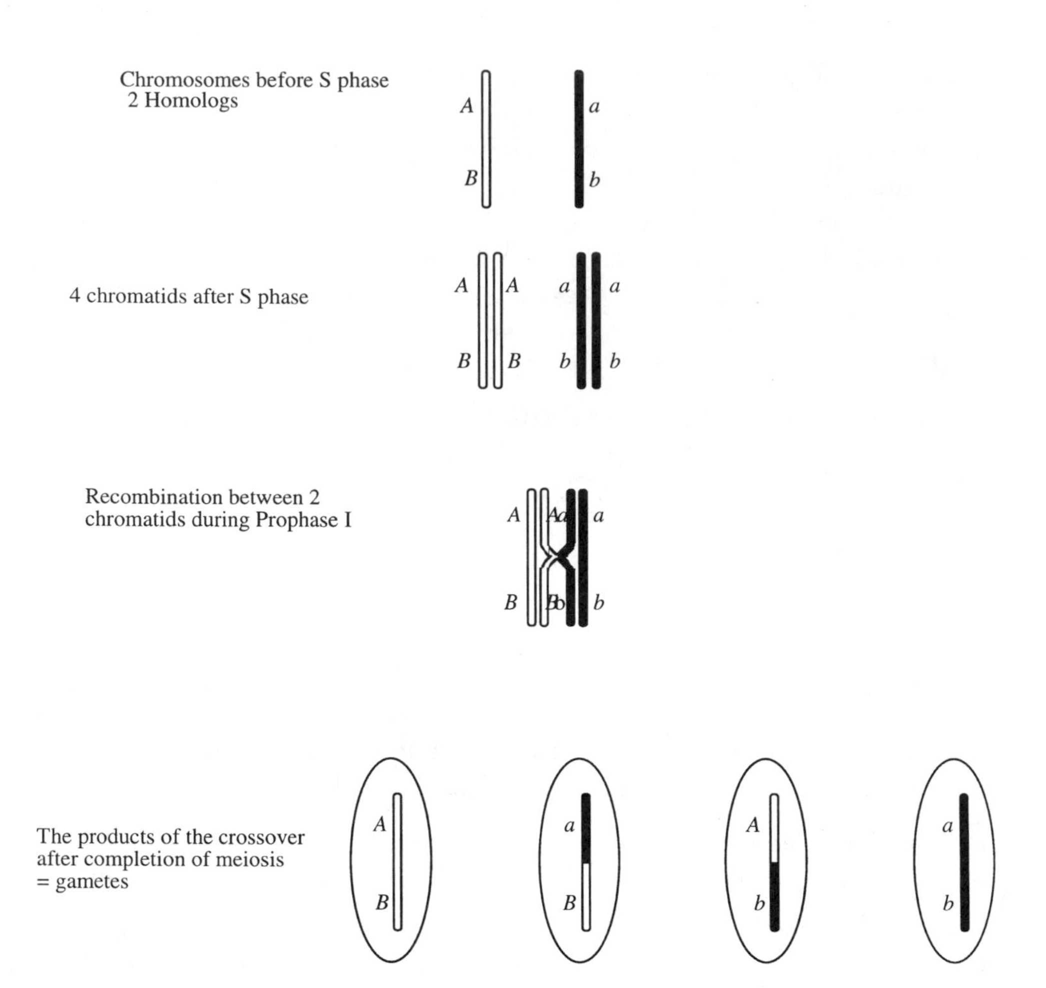

Figure 4.2 A crossover.

initiates meiotic recombination between two of the four chromatids that comprise the synapsed homologs. Some of the time, the repair will cause the two chromatids to exchange chunks of themselves. This exchange is **reciprocal**, meaning the swap of segments is equal, and it is both physical and genetic. In organisms whose chromosomes bear microscopically visible landmarks (or characteristics), the exchange is obvious in the reorganization of the landmarks. Also visible are distinctive X-shaped structures holding the homologs together from the end of prophase I through metaphase I; these structures, called **chiasmata** (**chiasma** is the singular), are thought to be sites of crossing over. Furthermore, in crosses where the chromosomes bear multiple heterozygous genes, the recombination is detected in the progeny as a reassortment of alleles, creating new combinations not possessed by the parents. This is a subtle, yet important point, so take note of it: Crossovers cannot be detected if they are not flanked by heterozygous genes, because our only genetic evidence of them is reassortment of alleles.

When Morgan detected progeny with nonparental allele configurations, he called the phenomenon **crossing over** and thought they were related to chiasmata. Crossing over is known to be necessary for proper disjunction of homologs during meiosis in nearly all organisms. (A notable exception is the male *Drosophila.*) Failure of homologs to cross over results in their missegregation at meiosis I (a phenomenon termed **nondisjunction**), eventually producing gametes with an incorrect number of chromosomes (see Chapter 7). Crossovers occur randomly and frequently, one to a few times per chromosome per cell (see also Topic 4). And with only two of the four chromatids per chromosome involved at each recombination site, the maximum frequency of recombination is 50% (2/4), even if every cell undergoing meiosis has a crossover between the two marker loci (i.e., heterozygous genes). But crossovers are finite in number; not every part of every chromosome recombines in every meiosis, so recombination frequency is less than 50% if the genes being analyzed are linked. How much less is the frequency? It depends on how far apart the genes are. As the distance between two genes increases, the frequency of recombination between those genes increases, up to 50%. Fifty percent is the maximum recombination frequency because 50% *is* random; even genes that are on different chromosomes will have 50% recombinant gametes, as shown in Topic 1. In Figure 4.2, the parent's alleles are arranged so that the dominant alleles are on one homolog and the recessive ones are on the other. This individual produces both of these parental gametes (*AB* and *ab*) as well as the recombined gametes (*Ab* and *aB*), although these latter gametes will be less frequent than the parental ones. The two reciprocal recombinants, so called because one crossover produces one of each recombinant-type progeny, are at equal frequency to each other, as are the two parental gametes. However, the ratio of parental and recombinant gametes depends solely on the distance between the two genes and can never be more than 1:1, because this ratio means independent assortment.

Some nomenclature is now necessary. The physical location of a gene is called a **locus** (plural, **loci**) and heterozygous genes are called **markers** because they enable detection of recombination, by changing allele configurations (or arrangements). There are two pairs of words used to describe allele configurations: ***cis*** and ***trans***, or **coupling** and **repulsion**. The *cis* (or coupling) configuration means the dominant alleles of both (all) genes are on the same homolog, for example, *AB* on one homolog and *ab* on the other. The *trans* (or repulsion) configuration means the dominant alleles are on opposite homologs, for example, *Ab* on one homolog and *aB* on the other. There is a useful way to represent this. Originally we wrote the dihybrid genotype as *AaBb*. If its parents are *AABB* and *aabb*, we write *AB*/*ab*, using a slash to separate one homolog's genotype from the other's. This configuration is *cis* (coupling). The parents would have to be different to result in the *trans* (repulsion) configuration: *Ab*/*aB*.

Topic Test 2: Linkage and Crossing Over

True/False

1. Linked alleles become unlinked by meiotic recombination.
2. A locus is a site on a chromosome.

Multiple Choice

3. A dihybrid has the *trans* configuration of alleles. How would we write its genotype to make this clear?
   a. *AaBb*
   b. *AB*/*ab*
   c. *Ab*/*aB*
   d. *Aa*/*Bb*
4. What are the genotypes of the parents of a dihybrid whose alleles are in *trans*?
   a. *AABB* and *AABB*
   b. *AaBb* and *AaBb*
   c. *AABB* and *aabb*

d. *AaBb* and *aabb*
e. *AAbb* and *aaBB*

Short Answer

5. What evidence of crossovers exists?
6. What is the evidence for linkage between genes in an experimental cross?

Topic Test 2: Answers

1. **True.** Provided the crossover (recombination) occurs between the two genes in question.
2. **True.** Usually this site is a gene, but it could be one of the visible structures mentioned also.
3. **c.** *Trans* means the dominant alleles are on separate homologs. The slash in the genotype is used to separate the homolog's alleles to make this distinction easier.
4. **e.** *Trans* alleles are *Ab/aB*. The genotypes in cross e are the only ones able to produce the *trans* dihybrid. Cross c could produce a dihybrid but its allele configuration would be *cis*, not *trans*.
5. The evidence of crossovers includes rearrangement (exchange) of cytological landmarks on physically marked chromosomes and new allele configurations in the progeny of multiply heterozygous individuals.
6. Evidence of linkage is the parental classes of progeny outnumbering the recombinant classes, instead of being equal as in independent assortment.

TOPIC 3: GENETIC MAPS

KEY POINTS

✓ *What is genetic distance?*

✓ *How is a genetic map constructed?*

✓ *How is a genetic map used to predict progeny?*

If crossover is proportional to distance, we can make maps of genes in their correct order with approximate distances between the genes. **Two-point mapping** is used to determine whether two genes are linked and, if they are, to determine the distance between them. Remember that for independent assortment, the testcross of a dihybrid will give four phenotypic classes of progeny in equal proportions (i.e., a 1 : 1 : 1 : 1 ratio). Linkage is in evidence if the ratio is not 1 : 1 : 1 : 1; that is, if the frequency of recombinant progeny is less than 50%. Practically speaking, the frequency of recombinant progeny should be about 40% or less to conclude that two genes are linked. This equivocation is wise because sample size can increase error and because as recombination frequencies increase, more crossovers will go undetected (see Topic 4).

As an example of linkage, consider Morgan's original cross demonstrating linkage of two autosomal genes in *Drosophila*. One gene in this cross controls eye color. Red, pr^+, is wild type and dominant; purple, *pr*, is recessive. The second gene determines wing length. Wild-type length, vg^+, is dominant. Vestigial wing length, *vg*, is recessive. (Aside: In *Drosophila*, as well as many other organisms, genes are named by the phenotype of the mutant and their symbols are one or more letters and/or numbers. A superscript plus sign [+] is often used to denote the wild type, and in *Drosophila*, capital letters indicate that the mutant alleles are dominant.) Morgan crossed a purple vestigial (doubly recessive) fly to a wild type to get heterozygous F_1 offspring. He then testcrossed a female F_1 fly to a doubly recessive male fly. (Since *Drosophila* chromosomes do not cross over during spermatogenesis, it is important that the heterozygote being testcrossed is female. This is not the case in most other organisms.) The 2,839 progeny of this cross are proportioned as follows (symbols indicate phenotype):

1,339 $pr^+ \ vg^+$

1,195 $pr \ vg$

151 $pr^+ \ vg$

154 $pr \ vg^+$

Since this is a testcross, we would expect a 1:1:1:1 ratio between these classes if the genes are not linked. Instead the ratio is roughly 8:8:1:1! The actual ratio is not important; it is different for each pair of genes. What is important (and is generally true when two genes are linked) is that the progeny fall into two groups, each with two phenotypic classes in a 1:1 ratio, and the ratio between the groups is unpredictable (but not 1:1). The groups are composed of reciprocal pairs of progeny, meaning that both members of the pair can come from the same meiosis in the dihybrid. Not by accident, the two most numerous classes, $pr^+ \ vg^+$ and *pr vg*, are parentals, and the two least numerous classes are recombinants.

We know the allele configuration of the female F_1 fly Morgan used in this testcross from two pieces of information. First, we know the F_1's parents. The recessive parent contributed *pr* and *vg* and the other parent contributed pr^+ and vg^+, so the dihybrid is *cis*. Secondly, we know she is *cis* from the greater abundance of *cis* progeny than of *trans* progeny in the testcross; the most numerous progeny classes are always parental. Morgan inferred that the abundance of these two parental classes was a result of the physical linkage between the two genes. Usually the particular allele configuration that came to the F_1 from her parents is transmitted unaltered to her progeny. Occasionally, however, the alleles appear recombined in the progeny, in a frequency proportional to the likelihood of recombination between them. Since this likelihood is related to the physical distance between them, we can use crossover frequency as the genetic distance between genes.

The orders of genes on the chromosomes are the same genetically and physically, but the distances are *not* identical. Genetic distances are measured in **map units** (**m.u.**), where 1 m.u. equals 1% recombination. Map units are also known as centimorgans (cM) in honor of T. H. Morgan. We can calculate the distance between the purple and vestigial genes in Morgan's experiment from the recombinant progeny classes, $pr^+ \ vg$ and $pr \ vg^+$. Sum together the number of progeny in these two classes, 151 + 154 = 305, and then divide by the total number of progeny, 305/2,839 = 0.107; this is the fraction of progeny that are recombinant. Multiply this number by 100 to get a percentage, 10.7%. This is equal to 10.7 m.u. Often this information is depicted as a **genetic map** in which a line represents a portion of a chromosome, dots or ticks on the line mark the positions of the mapped genes, and the distance is labeled, as shown on page 72. Note: This drawing is a map for *both* homologs and all possible allele configurations; all alleles of the purple gene, including the wild-type allele are 10.7 m.u. away from the vestigial locus. And this map is identical to the map that shows the vestigial gene on the left and the purple on the right [because it does not show a centromere or telomere (see Chapter 5)].

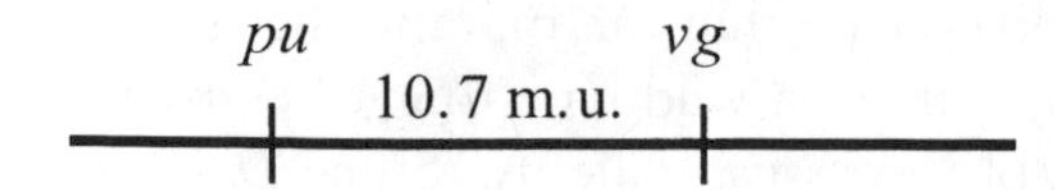

It is very likely that you will be required to calculate map distance from two-point mapping data and to do the reverse: calculate expected numbers of recombinants by type from a given map. As an example, this map tells us that the frequency of *pr vg* progeny from the testcross of a $pr^{+}vg/pr\ vg^{+}$ fly is 0.0535. The frequency of recombinant progeny is 0.107 from the map, and there are two types of recombinants. We only want one type (*pr vg*), so $0.107/2 = 0.0535$.

Topic Test 3: Genetic Maps

True/False

1. The genetic distance is equal to the physical distance between genes.
2. The numbers of parental and recombinant progeny are the same when linkage is observed.

Multiple Choice

3. The testcross of a dihybrid yields four classes of progeny. Which of the following are the phenotypes of reciprocal progeny?
   a. ab and ab
   b. ab and aB
   c. AB and Ab
   d. AB and AB
   e. Ab and aB
4. Which one of the following statements is false?
   a. Parentals are the most numerous progeny classes.
   b. The allele configuration of the dihybrid is the same as the chromosome passed on to the parental-type progeny.
   c. The parental-type progeny classes will be equivalent in number.
   d. The genetic distance is the frequency of crossover.
   e. Crossover frequency between genes decreases as distance increases.

Short Answer

5. The genes for normal (*O*) versus oblate (*o*) tomatoes and smooth (*P*) versus peach (*p*) tomatoes are 17 m.u. apart. A *cis* dihybrid for these two genes is testcrossed. What frequency of normal, smooth progeny should we expect?
6. A tomato that is heterozygous for a gene that makes stems purple or green and also for a gene that makes foliage hairy or hairless is testcrossed to a green hairless plant and the 500 progeny are as follows:

42 purple, hairy
202 purple, hairless
209 green, hairy
47 green, hairless

What is the distance between these genes?

Topic Test 3: Answers

1. **False.** Genetic distances and physical distances are not proportional.
2. **False.** These two classes are equal only when independent assortment is occurring.
3. **e.** By putting these two back together, *aB/Ab*, we get a dihybrid genotype, so these must be the reciprocal pair.
4. **e.** Distance and crossover frequency are proportional, because distance is determined directly from crossover frequency. As one increases, so does the other.
5. A *cis* heterozygote has the right allele configuration for the parental-type offspring to be normal and smooth. The frequency of parental types are 100 − 17 = 83%. This includes both parental types of progeny so dividing by 2 will give the frequency of just the normal, smooth type: 83/2 = 41.5%.
6. The heterozygote must be purple-hairless/green-hairy. The recombinants are purple, hairy and green, hairless, which total 42 + 47 = 89. Divide by 500 to get the frequency: 89/500 = 0.178. The map distance is 17.8 m.u.

TOPIC 4: GENE ORDER AND DISTANCE

KEY POINTS

✓ *How is three-point mapping carried out?*

✓ *What are double crossovers?*

There are two methods for determining the order of genes on a chromosome. The first method is the one described in Topic 3: Perform large numbers of two-factor crosses, using all the different combinations of genes to be placed on the map. The more efficient method is to perform fewer crosses using trihybrids instead of dihybrids. This is called **three-point mapping**; it gives distances and also establishes the order of three genes through a single cross instead of the three crosses that would be required for mapping with dihybrids. The experimental process is basically the same as that for two-point mapping: The trihybrid is testcrossed, and the progeny are classified as parental versus recombinant types. The difference is that there are many more classes of recombinants for a three-point cross.

Consider the *cis* trihybrid *ABC/abc* as an example. The testcross of this creature creates parental-type progeny of phenotype ABC or abc. In addition, there are six recombinant classes: those crossed over between *A* and *B* (*Abc* and *aBC*), those crossed over between *B* and *C* (*ABc* and *abC*), and those crossed over between *A* and *B and* between *B* and *C* (*AbC* and *aBc*). This last group

Table 4.1 Progeny of a Testcrossed Trihybrid Squirrel

CHARACTERISTICS	NO. OF PROGENY
long, thick, fast	142
short, sparse, slow	146
long, sparse, slow	10
short, thick, fast	12
long, thick, slow	34
short, sparse, fast	31
long, sparse, fast	310
short, thick, slow	315
total	1,000

represents **double crossovers**, that is, two crossovers that occurred close together in the same meiotic cell. Remember that crossovers are independent; therefore, we can calculate the frequency of two occurring together by multiplying the frequencies of either occurring alone (see Topic 1 in Chapter 2). Hence, double crossovers are much rarer than a single crossover in the same interval (i.e., between the same two genes). The benefit in recognizing this fact is that if the genes are linked, a quick scan of the progeny from a three-point mapping cross will reveal two reciprocal classes with many fewer members than the other six classes; these two classes are the products of double crossover. Again, provided the three loci are linked, the most numerous pair of classes will be the parentals. If one gene is not linked to the other two, there will be four classes sufficiently numerous to be parental, instead of two. This is one case when being able to recognize reciprocal classes is helpful. To analyze the progeny, identify the parentals, then identify the four recombinant classes that are crossed over for a pair of the three genes and determine the distance between them. Repeat this second step for the other two combinations of gene pairs.

Consider a hypothetical example in which the phenotypes of classes are given rather than the allele symbol. You will also see problems in which the allele symbols are given instead of the phenotypes. You should be able to solve both kinds of problems. This fictional three-point mapping cross concerns three characters of squirrels: tail length, fur thickness, and climbing speed. Long tails are dominant to short, thick fur is dominant to sparse, and fast climbing is dominant to slow. A trihybrid of unknown origin is repeatedly crossed to a homozygous recessive squirrel to generate the numbers of progeny listed in **Table 4.1**.

First, identify the parentals—long, sparse, fast and short, thick, slow—because these are the most numerous (note that they are also a reciprocal pair). This tells us the trihybrid is *cis* for tail length and fur, and climbing is *trans*. Knowing the allele configurations enables us to determine which interval is crossed over in the recombinant progeny. Next, we find the recombinants for each interval. Recombinants in the tail-fur interval are long, thick (142 + 34) and short, sparse (146 + 31), totaling 353. Divide by the total number of progeny (353/1,000 = 0.353) and multiply by 100 to get the distance (0.353 × 100 = 35.3 m.u.) from the tail locus to the fur locus. Repeating this procedure for the other two intervals, we get 31.0 m.u. for the distance between the speed locus and the fur locus and 8.7 m.u. for the distance between the speed locus and tail locus. To draw the map for these genes we need to determine the best arrangement for the loci, one that makes the most sense of the map distances and correctly predicts the double crossover class. One of the three distances (35.3 m.u.) appears to be the *rough* sum of the other two. This distance therefore measures the outside markers, tail and fur, and speed must be between them. The map for these genes looks like:

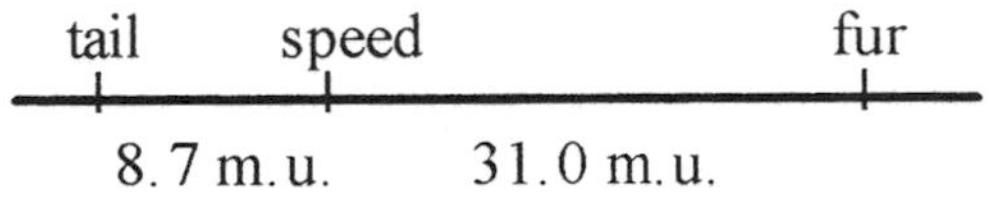

tail speed fur

8.7 m.u. 31.0 m.u.

This predicts that the double crossover class should be nonparental for the speed locus. One of the parentals is long, sparse, fast, so one of the double crossovers should be long, sparse, slow. Indeed it is one of the rarest classes, confirming the map order.

The last point to make about this map is that the distances are not exactly additive; the observed tail-to-fur distance (35.3 m.u.) is smaller than the calculated distance (8.7 + 31.0 = 39.7 m.u.). The reason for this discrepancy is the existence of multiple crossovers. The long, sparse, slow and short, thick, fast progeny are double recombinant. Because they appeared parental for the tail-fur interval, we did not count them in our calculation of that distance. We correct the map for these double crossovers by adding them into our calculation of that distance *twice*. Each one of the double recombinants arose from two crossover events in the interval, so they need to be doubled in number before adding to the other single recombinants: 10 + 12 = 22, then 22 × 2 = 44. Add to the single recombinants, 353 + 44 = 397, and divide by the total number of progeny to get 0.397, or 39.7 m.u. This distance equals the sum of the distances for the smaller intervals.

This added accuracy is another benefit of three-factor mapping. As distance between two genes increases, the likelihood of there being more than one crossover increases. While single crossovers between *A* and *B* in the *AB/ab* dihybrid are detected as recombinant progeny, two crossovers in the same interval fail to change the allele configuration. Progeny that have inherited such recombined chromosomes therefore are mistaken for parental types. This causes the distance between the two markers to be underestimated and the degree of underestimation is proportional to the size of the interval because crossover frequency increases with increasing distance. The only reason we were able to detect *any* double crossovers in the squirrel example is that there was a third, middle marker which enabled simultaneous detection of single crossovers to each side.

Topic Test 4: Gene Order and Distance

True/False

1. Double crossovers can only be detected if there is a heterozygous gene on each side of each crossover.
2. Double crossovers are nothing more than two single crossovers that occurred at the same time.
3. The distance between the outside markers in a three-point testcross will be overestimated.

Multiple Choice

4. The best method for determining gene order for an entire chromosome is
   a. many monohybrid crosses.
   b. many dihybrid crosses.
   c. many dihybrid testcrosses.
   d. many trihybrid crosses.
   e. many trihybrid testcrosses.

5. The rarest classes of progeny in a three-point testcross are
   a. the parental classes.
   b. the recombinant classes.
   c. the double crossover classes.
   d. the homozygous recessive class.
   e. none of the above.

6. Which of the following is not true of three-point testcrosses?
   a. They determine the frequency of double crossover.
   b. They determine the frequency of single crossovers.
   c. They determine the order of genes on a chromosome.
   d. They require a homozygous dominant parent.
   e. None of the above are false.

Short Answer

7. A breeder wants to map three genes in tomatoes that determine the appearance of the fruits and stems. The fruit color is controlled by alleles of a gene that make tangerine-colored (t) or red (T) fruits. Alleles of a second gene determine uniformly colored (u) or green-base (U) fruits. Alleles of a third gene cause stems to be hairy (h) or smooth (H). (This is a separate hairy gene from the one appearing in Topic Test 3, question 6.) The breeder testcrossed a trihybrid of unknown origin and obtained 500 progeny:

   151 red green-base smooth
   153 tangerine uniform hairy
   11 red green-base hairy
   13 tangerine uniform smooth
   42 red uniform hairy
   39 tangerine green-base smooth
   48 red uniform smooth
   43 tangerine green-base hairy

   Draw a linkage map, including distances, for these three genes.

Topic Test 4: Answers

1. **True.** Single crossovers are detected by the reassortment of two pairs of linked alleles, so double crossovers will require an additional pair of linked alleles, to create more new classes of progeny.

2. **True.** This is why we add two times the number of double crossovers to the outside marker distance to obtain a more accurate measure of that distance.

3. **False.** This distance will initially be underestimated because the double crossovers will not be counted in the calculation of this distance. The distance can be corrected after the gene order is determined so that it is the sum of the two smaller distances.
4. **e.** The trihybrid testcross (i.e., three-point mapping) is the most efficient method to determine gene order.
5. **c.** The double crossover progeny will be the rarest because they are created by the simultaneous occurrence of two rare events (single crossovers). The product rule tells us that these will be far less frequent than either of the single crossover recombinant classes.
6. **d.** One of the parents must be triply heterozygous and the other must be triply recessive for the same loci.
7. The parentals are red green-base smooth and tangerine uniform hairy. The distance between red/tangerine and green-base/uniform is given by the progeny that are red uniform and tangerine green-base (42 + 39 + 48 + 43 = 172), divided by the total, 172/500 = 0.344, or 34.4 m.u. The distance between green-base/uniform and smooth/hairy is given by the progeny that are green-base hairy and uniform smooth (11 + 13 + 48 + 43 = 115), divided by the total, 115/500 = 0.23, or 23 m.u. The distance between red/tangerine and smooth/hairy is given by the progeny that are red hairy and tangerine smooth (11 + 13 + 42 + 39 = 105), divided by the total, 105/500 = 0.21, or 21 m.u. These distances are consistent with a map that shows the hairy locus in the middle:

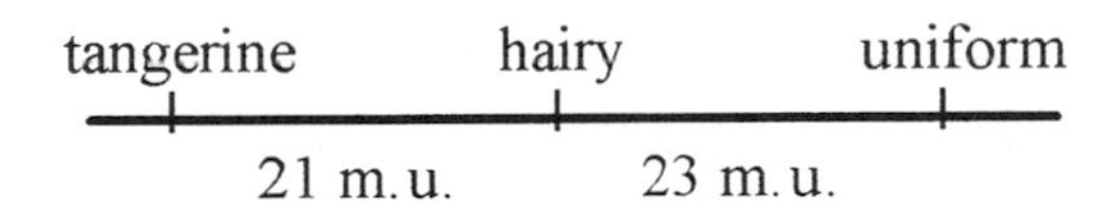

Double crossovers are red green-base hairy and tangerine uniform smooth. The interval these should be added to is tangerine to uniform: [172 + (24 × 2)]/500 = 0.44. Therefore, the distance between tangerine and uniform is 44 m.u., which agrees with the sum of the two smaller intervals.

TOPIC 5: INTERFERENCE

KEY POINTS

✓ *What is interference?*

✓ *How can interference be measured?*

It has been observed many times in many different experimental organisms that crossover within one region reduces the likelihood of a second crossover occurring in an adjacent region. This is called **interference**. This finding was a surprise at first because crossovers were believed to be independent events, and the basis for this phenomenon is still not understood. When we observe fewer double crossovers than we expect based on map distances derived from single crossovers, we say there is interference. Not all intervals will show interference, and some even show more double crossovers than expected. When we observe more double crossovers than we expected, we say there is **negative interference**.

Intervals that show interference or negative interference will consistently do so, and the amount of interference shown can be used to predict the progeny of future crosses. Interference and negative interference are quantifiable and are calculated from experimental data. The observed double crossover frequency, divided by the expected frequency of double crossovers, equals the **coefficient of coincidence, C**. Interference (I) is then calculated as I = 1 − C. For example, if gene *A* and gene *B* are separated by 10 m.u. and gene *B* and gene *C* are 2 m.u. apart, with the gene order being *ABC*, the expected double crossover frequency is $0.1 \times 0.02 = 0.002$. If, in the progeny of the cross *ABC*/*abc* × *abc*/*abc*, there are eight double crossovers (*AbC*/*abc* or *aBc*/*abc*) out of 10,000 total progeny, the observed frequency of double crossovers is 8/10,000 = 0.0008. C = 0.0008/0.002 = 0.4, so I = 1 − C = 1 − 0.4 = 0.6. We could say that interference was 60% in this cross or we could say that only 40% of the expected double crossovers occurred.

Notice that interference and coefficient of coincidence are inversely related. If fewer double crossovers than expected are observed, C will be less than 1, and I will be a positive number between 0 and 1; this is interference. When C equals 0 and I equals 1, interference is said to be complete because no double crossovers are observed. If the expected number of double crossovers are observed, C equals 1 and I equals 0; no interference is observed. If more double crossovers than expected are observed, C is greater than 1 and I is less than 0 (i.e., it is negative); this is negative interference. Negative interference is uncommon.

Topic Test 5: Interference

True/False

1. Interference is expected whenever the two intervals in the cross are of the same size.
2. Interference means fewer double crossover progeny than expected are observed.
3. The magnitude of interference changes over time, such that for particular intervals, the expected double crossovers will only be known within the context of each trihybrid cross.

Multiple Choice

4. The coefficient of coincidence is calculated by
   a. dividing observed double crossover frequency by expected double crossover frequency.
   b. dividing expected double crossover frequency by observed double crossover frequency.
   c. subtracting interference from C.
   d. adding interference to 1.
   e. calculating expected double crossover.
5. The distance between gene *A* and gene *B* is 20 m.u. and the distance between gene *B* and gene *C* is 5 m.u. (gene order is *ABC*). Out of 500 progeny of the testcross of a trihybrid, 3 exhibit double crossover. How much interference is observed?
   a. 0.06
   b. 0.4
   c. 0.6
   d. 0.94
   e. 0.98

Short Answer

6. Calculate the coefficient of coincidence and the value of interference for question 7 in Topic Test 4.
7. Calculate the coefficient of coincidence and the value of interference for question 8 in Topic Test 4.

Topic Test 5: Answers

1. **False.** Interference cannot be predicted. Some intervals consistently show interference, others do not, and any interval that has not been previously tested is not known to have or not to have interference.
2. **True.** The name *interference* can be remembered as follows: One crossover interferes with the formation of a second crossover in the same region.
3. **False.** For a particular interval, the magnitude of interference is constant and can be used for predicting progeny from crosses.
4. **a.** Coefficient of coincidence is equal to observed double crossover frequency divided by expected double crossover frequency.
5. **b.** Expected double crossover frequency is $0.2 \times 0.05 = 0.01$. Observed frequency of double crossover is $3/500 = 0.006$. $C = 0.006/0.01 = 0.6$, so $I = 1 - 0.6 = 0.4$.
6. Expected double crossover frequency is $0.21 \times 0.23 = 0.048$. Observed frequency of double crossover is $24/500 = 0.048$. $C = 0.048/0.048 = 1.0$, so $I = 1 - 1.0 = 0.0$. There was no interference.
7. Interference is not relevant to this problem because one of the loci was unlinked, so there was no way to detect double crossovers.

IN THE CLINIC

Linkage of genes can be helpful in determining the existence and locations of genes that contribute to human disease. For the first half of this century, various investigators collected pedigrees for families afflicted by two X-linked recessive disorders, hemophilia and color blindness. Many pedigrees showed clear evidence of linkage. Now we know that there are two hemophilia genes on the X chromosome, one of which is sufficiently close to the color blindness locus to show linkage in pedigrees. Now linkage is used to identify human disease genes that are near physical loci (see Chapter 9, Topics 3 and 5).

DEMONSTRATION PROBLEM

See the trihybrid testcross examples in Topics 3 and 4. Make a genetic map from two-factor testcross data. Make a genetic map from a three-factor testcross data. Determine frequency of one genotype from a genetic map.

Chapter Test

True/False

1. The location of recombination is random.
2. Linkage is not observed in the progeny of a selfing.
3. *AB*/*ab* is the *trans* configuration.
4. Map distances are strictly additive.
5. Recombinant progeny come from one crossover each.
6. Double crossover is the simultaneous occurrence of two single crossovers.
7. Recombination is what unlinks genes.
8. The most numerous classes of progeny in testcrosses are the parental types.

Multiple Choice

9. The least numerous class of progeny resulting from the testcross of a trihybrid is/are
   a. double crossovers.
   b. single crossover in the smallest interval.
   c. single crossover in the largest interval.
   d. parental type.
   e. recombinant type.
10. What prevents map distances from being exactly additive?
    a. Positive interference
    b. Negative interference
    c. Multiple crossover
    d. Linkage
    e. None of the above
11. A testcross of a furry brown heterozygote to a bald pale recessive homozygote produces 179 furry pale, 168 bald brown, 38 furry brown, and 44 bald pale offspring. What is the map distance between these two genes?
    a. 0.191 m.u.
    b. 19.1 m.u.
    c. 26.2 m.u.
    d. 82 m.u.
12. The genes *Black* and *Slick* are 23 m.u. apart and genes *Slick* and *Curly* are 14 m.u. apart. (The order is *Black-Slick-Curly*.) A trihybrid for these genes was testcrossed and it gave 26 double crossover progeny out of a total of 1,000. What is the value for interference in this experiment?
    a. 1
    b. 0.807
    c. 0.193
    d. 0.026
    e. 0

13. Testcrosses of dihybrids give what ratio of progeny?
    a. 1 : 1 : 1 : 1 if the genes are linked
    b. 1 : 1 : 1 : 1 if the genes are not linked
    c. 9 : 3 : 3 : 1 if the genes are linked
    d. 9 : 3 : 3 : 1 if the genes are not linked
    e. None of the above

14. If I equals 1,
    a. no double crossovers are observed.
    b. interference is complete.
    c. more double crossovers than expected are observed.
    d. both a and b are true.
    e. both b and c are true.

15. Two genes are 15 m.u. apart on the chromosome. What is the frequency of the double recessive progeny from a selfing of a *trans* dihybrid?
    a. 0.15
    b. 0.075
    c. 0.023
    d. 0.0056
    e. 0.0025

Short Answer

16. What is the clearest indication that two genes are not linked?

17. What is crossover, exactly?

Essay

18. Provide a *complete* interpretation of the following progeny set, which came from a single cross.

    201 glossy
    438 fragile, glossy, red
    197 glossy, red
    427 fragile, glossy
    431 wild-type
    434 red
    189 fragile, red
    205 fragile

19. Describe what each of the classes of progeny from a testcrossed trihybrid tell you about linkage. Assume all three of the trihybrid's heterozygous genes are linked.

Chapter Test Answers

1. **T** 2. **F** 3. **F** 4. **F** 5. **F** 6. **T** 7. **T** 8. **T** 9. **a** 10. **c** 11. **b** 12. **c**
13. **b** 14. **d** 15. **d**

16. In the testcross of the double heterozygote for these genes, the fraction of recombinant progeny equals the fraction of parental progeny. Put another way, the recombinants are 50% of the total progeny.

17. Crossover is the physical exchange of chromosomal segments between homologs during meiosis. It serves to unlink genes that lie on the same chromosome.

18. There are eight progeny classes and they fall into four groups of reciprocal phenotypes. This implies that the parents were a trihybrid and a triple recessive homozygote. The 438 fragile glossy red and 431 wild-type progeny appear to be reciprocal classes, as do the 427 fragile glossy and 434 red, 189 fragile red and 201 glossy, and 197 glossy red and 205 fragile progeny. It appears that there are four parental classes and four recombinant classes, consistent with independent assortment of one gene. The four parental classes consist of two that are fragile and glossy and two that are neither (i.e., they are wild type). Therefore the trihybrid parent must have been *cis* for the fragile and glossy alleles. The red gene is independently assorting; evidence for this conclusion is found in the failure of red to segregate consistently with fragile and glossy in the parental classes. In fact, this is why there are four parental classes. The four recombinant classes reflect the distance between fragile and glossy. The sum of the recombinant classes is 201 + 197 + 189 + 205 = 792. The map distance is (792/2,522) × 100 = 31.4 m.u.

19. Of the eight progeny classes expected, the reciprocal pairs of classes will aid in identifying the allele configuration in the heterozygous parent, the intervals showing crossover, and the gene order. The parental classes will be the two most numerous classes and they will indicate the allele configuration in the heterozygous parent. This knowledge allows the recombinant classes to be identified by which interval is showing crossover. For each pair of genes, two of the three pairs of reciprocal recombinants will be used for the map distance calculation. Draw a map that is consistent with the three calculated map distances. The two double crossover classes will be the rarest progeny and their phenotypes should agree with the drawn map: The gene for which they are nonparental should be the middle gene on the map. Interference is calculated from the observed frequency of double crossover and the expected frequency based on the calculated map distances.

Check Your Performance:

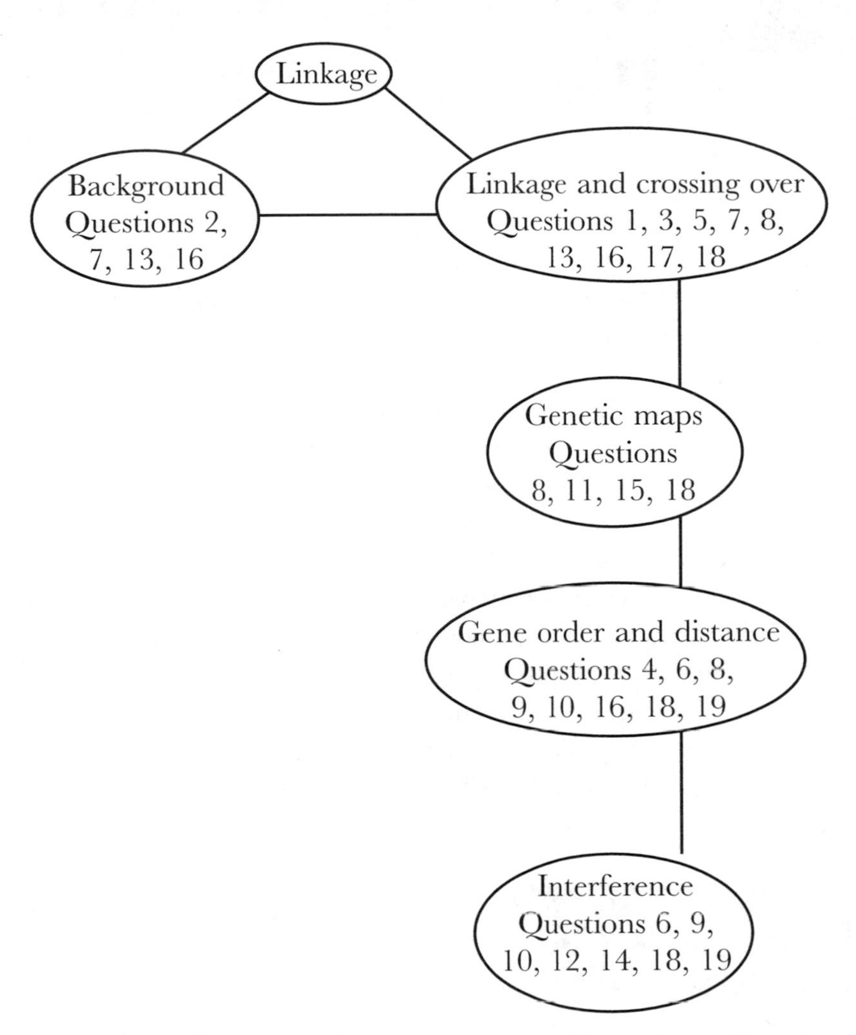

Use this chart to identify weak areas, based on the questions you answered incorrectly in the Chapter Test.

DNA, RNA, and Proteins

From the time of Mendel's discovery until the mid 1940s, no one knew the chemical basis of heredity. One popular theory was that protein, a significant component of chromosomes, was the heritable material. In the end, it took two excellent experiments to convince the skeptics that **DNA (deoxyribonucleic acid)** is the genetic material. Most organisms use DNA as their genetic material, but some viruses use **RNA (ribonucleic acid)** instead of DNA. The structure of DNA was determined in the 1950s, and it suggested the way in which DNA is replicated. The information contained within DNA is copied into RNA molecules by a process called **transcription**. Another process, **translation**, synthesizes proteins based on the sequences of RNA molecules. In the years since these discoveries, interest in the entire genetic complement of organisms has boomed. One complete copy of the DNA in an organism is defined as that organism's **genome**. That is, the genome of a diploid organism is its haploid set of chromosomes and all of the genes contained therein. The study of genomes addresses many diverse questions, such as how and which gene products interact in a single organism and how diverse organisms are related to each other.

ESSENTIAL BACKGROUND

- Basic cellular structure
- Chromosomal theory of inheritance (Chapter 1)
- Basic chemistry (elements, classes of chemicals, radioisotopes, hydrogen bonding)
- ATP as energy source for cells

TOPIC 1: DNA STRUCTURE

KEY POINTS

✓ *What is the structure of DNA?*

✓ *What are the chemical components of DNA?*

The structure of DNA was proposed in 1953 by J. Watson and F. Crick. Data from two sources were critical to their model. E. Chargaff had determined an empirical relationship between the chemicals that comprise DNA (see below). R. Franklin and M. Wilkins collected x-ray diffraction patterns from a DNA fiber, which revealed a helical structure and some of its dimensions. As Watson and Crick proposed, DNA usually exists as two strands, twisted together into a right-handed helix, called the **double helix**. Each strand is a polymer of repeating units called **nucleotides**. A generic nucleotide (**Figure 5.1**) has three components: a phosphate, a sugar

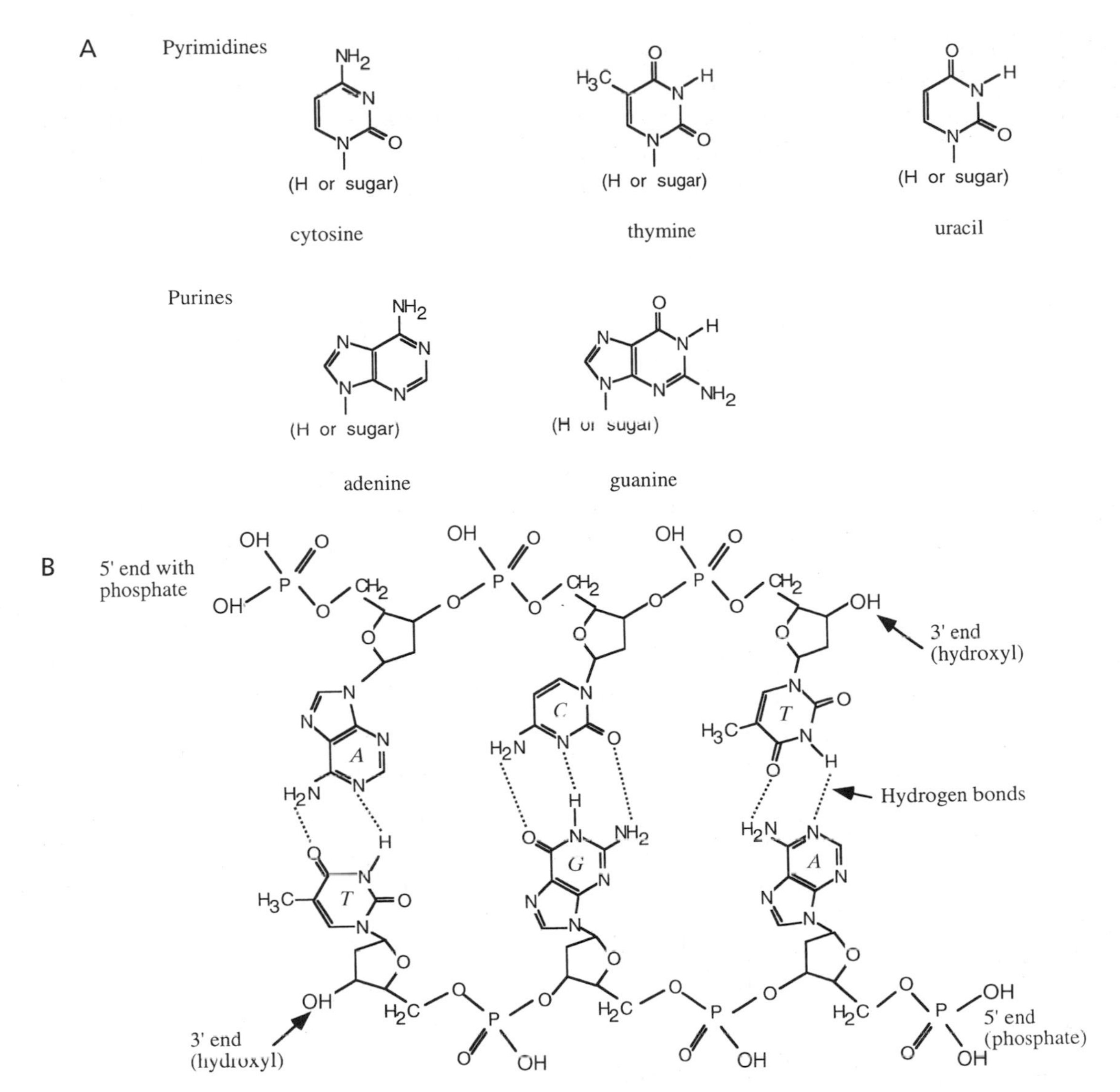

Figure 5.1 A. The structures of nitrogenous bases. B. Three base pairs of double-stranded DNA.

(deoxyribose), and a nitrogenous base. The abundant phosphate groups (one per nucleotide) give DNA a negative charge at physiological pH. Within DNA, deoxyribose exists as a five-member ring in which carbon atoms 1 through 4 are part of the ring, along with an oxygen atom from the fourth carbon's hydroxyl group. The fifth carbon atom sticks out from the ring. To distinguish them from other parts of the nucleotide and from each other, the carbon atoms are numbered 1′ ("one prime") through 5′ ("five prime"). In a nucleotide, the base is attached to the 1′-carbon atom of the sugar; the phosphate is attached to the 5′ carbon atom. In the DNA strand (see Figure 5.1b), adjacent nucleotides are bonded together at one nucleotide's 3′ carbon atom and the second nucleotide's phosphate. These linkages form the "backbone" of the DNA molecule. Because the backbone is not symmetrical, the entire DNA strand is polar. The *5′ end* of the strand is a phosphate, attached to the 5′ position of deoxyribose. The *3′ end* of the strand is the only deoxyribose in that strand whose 3′ carbon is not bonded to a phosphate. The two strands in the double helix have opposite polarity (they are **antiparallel**); the 5′ end of one strand is near the 3′ end of the other strand (see Figure 5.1b).

The central portion of the double helix is occupied by stacked pairs of nitrogenous bases. The nitrogenous bases found in DNA come in four varieties that belong to two chemical classes (see Figure 5.1a). **Adenine (A)** and **guanine (G)** are **purines**; **cytosine (C)** and **thymine (T)** are **pyrimidines**. **Uracil (U)** is an additional pyrimidine found in RNA in place of thymine. These names and abbreviations are also used for the nucleotides, although the correct names are deoxyadenosine 5′-monophosphate (or dAMP), deoxyguanosine 5′-monophosphate (or dGMP), deoxycytidine 5′-monophosphate (or dCMP), and deoxythymidine 5′-monophosphate (or dTMP).

The two strands in a double helix are held together by interaction between the nitrogenous bases, which exhibit specific hydrogen bonding such that A in one strand pairs with T in the opposite strand and C pairs with G. The A:T base pairs are held together by two hydrogen bonds and the C:G base pairs are held together by three hydrogen bonds (see Figure 5.1b). Because of the specificity of base pairing, the sequence of one strand determines the sequence of the other. For this reason, the two DNA strands are said to be **complementary**. Note that "complementary" is very different from "identical": In a complementary pair of strands, A bonds with T, not with A. E. Chargaff showed that regardless of the species from which it was isolated, double-stranded DNA contains equal amounts of A and T bases and equal amounts of G and C bases, and this information helped elucidate the double helical nature of DNA (see above). This relationship among the bases is true because of the specific base pairing within DNA double helices. Chargaff also determined that the amounts of A plus T versus G plus C vary between species. These two observations have come to be known as **Chargaff's rules**.

RNA is used as the genetic material in some viruses. In all other organisms it serves in other functions (see Topic 4). RNA is usually single stranded and quite similar to a single strand of DNA with two exceptions. First, the sugar moiety is ribose, not deoxyribose. The sole difference is the presence of a hydroxyl group on the 2′ carbon atom of ribose and its absence in deoxyribose (a hydrogen atom is present, instead). Secondly, the pyrimidines present in RNA are cytosine and uracil, not thymine. In double-stranded RNA, uracil base pairs with adenine.

Topic Test 1: DNA Structure

True/False

1. Double-stranded DNA has bases on the outside of the helix.
2. DNA strands exhibit polarity.
3. The phosphate group is at the 5′ end of the DNA molecule.

Multiple Choice

4. Which of the following statements does not follow from Chargaff's rules?
   a. The amounts of A and T are the same.
   b. The amounts of C and G are the same.
   c. The amount of A plus T may not be the same as the amount of C plus G.
   d. A and T (on opposite strands) form one base pair.
   e. A, G, C, and T attach to the sugar moiety.
5. The nucleotides in DNA are composed of all but which of the following?
   a. Deoxyribose
   b. Adenine

c. Cytosine
d. Uracil
e. Phosphate

Short Answer

6. Name the people who directly contributed to the determination of DNA's structure and summarize their contributions.
7. What does it mean to say that DNA strands are complementary?

Topic Test 1: Answers

1. **False.** Double-stranded DNA has the phosphate-sugar backbone on the outside of the helix and the bases on the inside.
2. **True.** The polarity arises because deoxyribose is polar and is a component of the DNA backbone.
3. **True.** The phosphate is attached to the 5′ carbon atom of deoxyribose in free nucleotides, so it forms the 5′ end of the strand.
4. **e.** Although this statement is true, it is not one of Chargaff's rules, which only address base composition. Answer d is the physical basis for Chargaff's rules.
5. **d.** Uracil is a pyrimidine found in RNA, not DNA.
6. E. Chargaff determined the base composition. Amounts of A equal amounts of T; amounts of C equal amounts of G. Amounts of A and T (and therefore, C and G) are not the same in all organisms. R. Franklin and M. Wilkins provided the evidence that DNA was a helix and two of its dimensions. J. Watson and F. Crick devised a model from these data and the knowledge of the components (phosphate, deoxyribose, and four bases).
7. Two strands that are complementary can base pair with each other. When they are aligned (i.e., antiparallel), adenines in one strand base pair with thymines in the opposite strand. Likewise, cytosines in one strand base pair with guanines in the other strand.

TOPIC 2: ORGANIZATION OF EUKARYOTIC CHROMATIN AND CHROMOSOMES

KEY POINTS

✓ *What are the dimensions of the double helix?*

✓ *What is the organization of DNA in the eukaryotic chromosome?*

✓ *What structural features are required for chromosome integrity?*

A **base pair (bp)** is a unit of physical distance; larger units are the **kilobase pair (kb)**, which equals 1,000 bp, and the **megabase pair (Mb)**, which is 1 million bp.

The double helix is a constant 2 nm in diameter because base pairs are composed of one purine and one pyrimidine each. One complete helical turn is only 3.4 nm long. This turn includes

10 bp, making each base pair 0.34 nm thick. Double-stranded DNA also has two grooves, the **major** and **minor** grooves, that twist around its exterior. One groove is larger than the other because the two sugar-phosphate backbones are unevenly spaced owing to the base pairing. Proteins that recognize and bind specific DNA sequences do so in the grooves.

Normally the DNA inside cells is intimately associated with chromosomal proteins. A eukaryotic chromosome is a single molecule of DNA, tightly compacted at metaphase into a highly ordered, flexible rod and less tightly organized during interphase. DNA is complexed with proteins (histones, for example) that provide structural organization, as well as regulation of replication and gene expression (see later topics). A variety of analyses have led to the following partial description of the levels of packing of the DNA duplex.

$$\text{DNA} \xrightarrow{\text{histones}} \text{10-nm fiber} \longrightarrow \text{30-nm fiber} \rightarrow \rightarrow \rightarrow \text{chromatin \& chromosomes}$$

Two loops, or 146 bp, of DNA is wound around an octamer formed from **histone** proteins H2A, H2B, H3, and H4 to compose a **nucleosome** whose diameter is 10 nm. Nucleosomes are linked by 25-bp stretches of DNA bound by a fifth ***linker*** histone, H1. Another 30 bp is naked between nucleosomes. Thus, one nucleosome occurs about every 200 bp. This **nucleofilament** is further condensed into a 30-nm-diameter fiber, and additional poorly understood compaction yields the chromatin and chromosomes that exist in living cells. One notable difference between chromatin and chromosomes is the presence of the enzyme **DNA topoisomerase II** in chromosomes, which causes negative supercoiling of DNA.

Two structural elements are important for the function of the chromosome: the **centromere** and the **telomeres**. The constrictions that appear on each metaphase chromosome in most species are centromeres. There is one centromere on each chromosome, and it is located in a characteristic position for that chromosome in that organism. In yeast, centromeres consist of three elements encompassing about 125 bp that are bound by specific proteins. Other organisms can have very different centromeres, as judging from the few others characterized. In meiosis and mitosis, spindle fibers attach to the centromere of each chromosome and hold the sister chromatids together until anaphase. Centromeres are required for proper segregation of the chromosome.

Telomeres are the physical ends of the DNA molecule. The DNA of the telomere is a simple sequence (5′-TTAGGG-3′ in humans) repeated many times and bound by special proteins. Its function is to protect the ends of the linear DNA molecule. When telomeres are experimentally removed from a chromosome in yeast, the chromosome gets progressively shorter with each cell division. The enzyme **telomerase** adds more repeats to the ends that already have repeats, thus preventing the shortening that arises from incomplete replication at the ends (see Topic 4).

Finally, a typical organism's DNA can be divided into two types. Some sequences are present in one or a few copies per cell. These are called **single copy sequences** and consist of genes and rare structural sequences. The second type is called **repetitive sequence**, with anywhere from a few to more than 100,000 copies per cell. These can be members of gene families, or noncoding functional sequences (e.g., those surrounding centromeres or telomeres and telomeres themselves), or sequences that have no known function. The repeats can be clustered at single sites, as for globin genes and ribosomal DNA genes, or can be scattered throughout the genome, as for Alu elements in some primates which have no known function.

Topic Test 2: Organization of Eukaryotic Chromatin and Chromosomes

True/False

1. Centromeres are required for chromosome segregation.
2. Histones are specific to condensed metaphase chromosomes.

Multiple Choice

3. What is the diameter of the nucleosome?
   a. 2 nm
   b. 3.4 nm
   c. 10 nm
   d. 30 nm
4. Why is the double helix a consistent diameter?
   a. It has major and minor grooves.
   b. Purines pair with pyrimidines.
   c. Hydrogen bonding is constant.
   d. The strands are polar.
   e. It winds around nucleosomes.

Short Answer

5. Summarize packaging of the eukaryotic DNA.
6. What are the dimensions of the double helix?

Topic Test 2: Answers

1. **True.** In most organisms, centromeres are the structures that are pulled to opposite poles at anaphase. Chromosomes that lack centromeres are not correctly separated, resulting in aneuploidy (see Chapter 7).
2. **False.** DNA is complexed with histones in chromatin as well as in chromosomes.
3. **c.** The chain of nucleosomes is the 10-nm fiber seen in the electron microscope.
4. **b.** Base pairing determines the diameter, since the backbone is not variable. Pyrimidines are small and they base pair with purines, which are larger. Neither class base pairs with itself so the DNA has a constant diameter.
5. The double helix is wound almost twice around histone octamers (two each of H2A, H2B, H3, H4) to make a 10-nm fiber. This fiber is packaged into a 30-nm fiber and then into other forms like the metaphase chromosome. The more condensed chromosome form requires additional proteins, such as DNA topoisomerase II.
6. The double helix is 2 nm in diameter and 3.4 nm high for every complete turn of the helix (10 bp).

TOPIC 3: REPLICATION

KEY POINTS

- ✓ *How is DNA replicated?*
- ✓ *What enzymes are involved?*
- ✓ *What are the activities of those enzymes?*

In every S phase preceding mitosis or meiosis, a cell makes a complete copy of its genome. The process is said to be **semiconservative** because each strand of the original double helix is a template for the new synthesis, so that two new double helices are formed. In each, one complete strand is old and one is new. This process is called **DNA replication** and the following description of it is based on in vitro studies of *Escherichia coli* enzymes and DNA molecules. **Figure 5.2** illustrates some of the more visually appreciated concepts described here. Replication begins at specific sequences (**origins**) in the DNA. There is one origin for the entire prokaryotic genome, which is a circular DNA molecule. The typical eukaryotic genome is organized into multiple chromosomes, each having numerous origins. Synthesis is initiated in *E. coli* by the binding of an initiator protein to the origin, followed by the unwinding (**denaturing**) of the double helix at this site by the enzyme **DNA helicase**. Pulling the strands of DNA apart is energetically unfavorable; for this reason, the helicase hydrolyzes ATP as it works.

The helicase recruits a second enzyme, **DNA primase**, to the site, which synthesizes a short (about 11 nucleotides) *complementary* RNA at the origin, using the DNA as the template. The RNA primer is absolutely required for the subsequent synthesis of a DNA strand because no known **DNA polymerase** (the enzyme that synthesizes DNA) is capable of initiating DNA synthesis without a 3′ end (RNA or DNA) from which to start. The primase and the polymerase are polar in their activities; the primase begins at what will be the 5′ end of the new strand and pro-

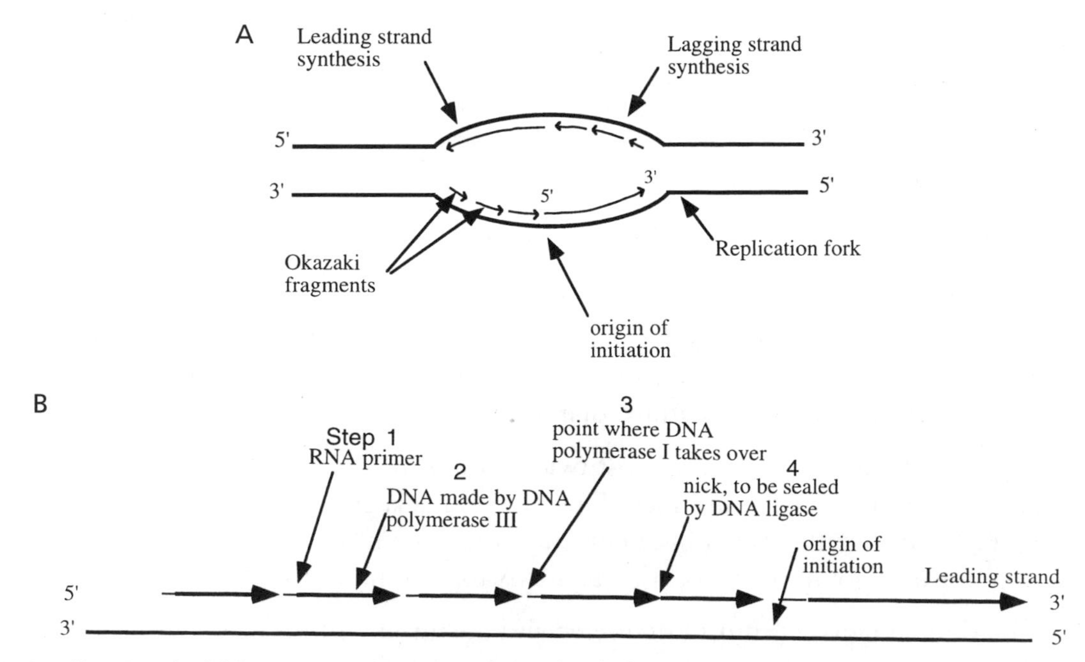

Figure 5.2 A. Replication bubble. B. Lagging strand synthesis in greater detail.

ceeds in the 3′ direction. The polymerase adds nucleotides only to the 3′ end of the primer and not to the 5′ end. Synthesis proceeds bidirectionally; at the same time that a primer and new strand are being synthesized on one of the template strands, the other original strand is also supporting the same activities. Each new strand that begins at the origin and proceeds in the 3′ direction is a continuous strand (the **leading strand**). There is a logistical problem at the sides opposite the leading strands. Since DNA polymerase only synthesizes in the 5′ to 3′ direction, the primer at the origin will not prime synthesis of anything 5′ of itself. Synthesis of this **lagging strand** is different from the synthesis of the leading strand, as described later. Both template strands support both leading strand and lagging strand synthesis (see Figure 5.2).

Note that DNA polymerase and primase do not add nucleotides randomly; the sequence of the primer and the new DNA strand is dictated by the template. Recall that RNA and DNA are polymers of nucleotides. The precursors in their synthesis are **nucleotide triphosphates**, which have a chain of three phosphate moieties where a single phosphate exists in the DNA or RNA. Primase and polymerase liberate the two "extra" phosphates as pyrophosphate, supplying the energy needed to add each nucleotide to the growing strand. In the next phase of the leading strand's synthesis, the multiprotein complex **DNA polymerase III** binds to the RNA primer and begins to synthesize a DNA that is attached to the primer and is complementary to the unwound template. The helicase continues to denature DNA ahead of the advancing polymerase, and the point at which the unwinding occurs is the **replication fork**. The extended structure of denatured strands and new synthesis is a **replication bubble**.

Replication of the lagging strand is different from that of the leading strand in several ways. First of all, the synthesis of the lagging strand is discontinuous. As more single-stranded DNA is created by the helicase, the primase repeatedly makes primers along the length of the lagging strand, about every 100 to 1,000 bp. DNA polymerase III extends each of the primers up to the next primer, and then dissociates from the DNA. At this point, a different polymerase, **DNA polymerase I**, takes over; it degrades the RNA primer from the junction, and finishes extending the DNA fragment up to the next newly synthesized fragment, leaving a nick (lack of covalent attachment) between the two fragments. These fragments are known as **Okazaki fragments**. The nick is sealed by the action of the enzyme **DNA ligase**, which also binds the first Okazaki fragment to the leading strand.

Lastly, there are a few additional players in the replication process. First, **single-stranded binding protein (SSB)** binds to and stabilizes the single-stranded regions created by helicase. These parts of the template are prone to **reannealing** (coming back together). SSB prevents reannealing while leaving the template available to the DNA polymerase. Secondly, the histones that normally package *eukaryotic* DNA are coordinately regulated so that as a cell begins S phase, it begins to make the large number of histones that will be needed to package all of the DNA for both daughter cells. Finally, a special polymerase, **telomerase**, resynthesizes the physical ends of the chromosomes in eukaryotes. This polymerase has its own template (an RNA), which is complementary to the repeating unit of the telomere. DNA polymerase III can synthesize the leading strand to the chromosome's end. It cannot complete the lagging strand because of the primer requirement: DNA polymerase cannot replace the primer made at the end of this template. With each successive replication, the discontinuously synthesized strand is at least 11 nucleotides shorter (because of the primer), unless something increases or restores the telomere length between rounds of replication. Telomerase adds more telomere repeats to the 3′ end of the "original" strand, effectively lengthening the chromosome.

Topic Test 3: Replication

True/False

1. Both primer and newly synthesized DNA are complementary to the old strand (i.e., the template).
2. The lagging strand is discontinuously synthesized.

Multiple Choice

3. Which of the following statements is incorrect?
   a. DNA polymerase I finishes synthesizing the Okazaki fragments.
   b. DNA polymerase III synthesizes from the primer.
   c. DNA primase and both polymerases synthesize a new strand in the 3′ to 5′ direction.
   d. DNA polymerases cannot begin synthesis without a primer.
   e. DNA polymerases proofread to prevent the incorporation of incorrect nucleotides.
4. Which one of the following is not necessary for prokaryotic DNA replication?
   a. DNA ligase
   b. DNA polymerase III
   c. DNA primase
   d. SSB
   e. Telomerase
5. Which one of the following is not true of DNA helicase?
   a. It unwinds the double helix.
   b. It prevents the reannealing of the unwound DNA.
   c. It allows DNA primase to make primers.
   d. It hydrolyzes ATP.
   e. It is found at the replication fork.

Short Answer

6. Make a list of the proteins involved in replication and note their purposes.

Topic Test 3: Answers

1. **True.** Replication is semiconservative because both of the old strands are the templates for the newly synthesized strands. Primase and the polymerases require a template so that they know what nucleotide to add.
2. **True.** Okazaki fragments are the products of discontinuous synthesis. They will be ligated together as replication proceeds.
3. **c.** Primase and both polymerases synthesize a new strand in the 5′ to 3′ direction.
4. **e.** Prokaryotes have circular genomes and therefore do not have telomeres. Eukaryotes have linear chromosomes and require telomerase to prevent their shortening.
5. **b.** This is the job of SSB.

6. Initiator protein recognizes the origin. DNA helicase unwinds the double helix. Single-stranded binding protein (SSB) stabilizes the unwound DNA. DNA primase synthesizes an RNA primer. DNA polymerase III synthesizes DNA beginning at the RNA primer. In lagging strand synthesis, DNA polymerase I synthesizes the rest of the Okazaki fragment, beginning at the DNA end where DNA polymerase III stopped and continuing until it abuts the next DNA strand, degrading the intervening RNA primer. DNA ligase seals the nick left between the end of this fragment and the beginning of the adjacent one. Telomerase adds more repeats onto the 3′ ends of linear DNAs.

TOPIC 4: RNA AND TRANSCRIPTION

KEY POINTS

✓ *What are the characteristics of RNA polymerases?*

✓ *What are the characteristics of RNA transcripts?*

A two-step process decodes the genes in DNA into proteins, according to the **central dogma**. First, DNA is copied into RNA (**transcription**). Second, the RNA message is converted into protein (**translation**). RNA is made from ribonucleotide triphosphates (the sugar moiety is ribose). **RNA polymerases** synthesize an RNA molecule that is complementary to one strand of the DNA (called the **template strand**) and thus is identical in sequence, except for the replacement of thymidine by uracil, to the other strand (called the **coding** or **nontranscribed strand**). Thus, RNA polymerases are DNA dependent and only a single strand of DNA is copied, not both. Prokaryotes have a single type of RNA polymerase that transcribes all of the organism's genes. Most of what is known about the process of transcription comes from studying the prokaryotic RNA polymerase.

The prokaryotic RNA polymerase is a complex of five proteins. Transcription in *E. coli* requires several specific DNA sequences within the gene. The polymerase copies the template strand, beginning at a specific DNA sequence, the **promoter**, and ending at another sequence, the **transcriptional terminator**. Transcription begins when the entire five-subunit complex of RNA polymerase binds to the promoter at the 5′ end of a gene. The **sigma** subunit leaves the complex after transcription has commenced. The typical *E. coli* promoter consists of the sequence TTGACA located about 35 nucleotides from the transcriptional start site (the **−35 box**) and the sequence TATAAT located about 10 nucleotides from the start site (the **−10 box** or **Pribnow box**). After binding the promoter, RNA polymerase denatures the DNA duplex in a small region surrounding the promoter and transcriptional start site, creating a "bubble." This bubble moves with the polymerase as it progresses down the template, and the DNA strands spontaneously reanneal after the polymerase passes through. Unlike DNA polymerases, RNA polymerases do not require a primer to initiate synthesis but, just as for DNA polymerases, nucleotides are added in the 5′ to 3′ direction. Elongation is the period of time when the polymerase is adding nucleotides to the growing chain; it continues until the polymerase has copied a termination sequence. A typical prokaryotic terminator consists of a self-complementary sequence so that the nascent RNA will hybridize to itself, forming a hairpin structure, followed by a series of T nucleotides in the template. A second type of terminator lacks the T string and is terminated in the presence of a termination protein named **rho (ρ) factor**.

In prokaryotes, there are three **ribosomal RNAs** (**rRNAs**; components of the ribosome) that are transcribed as a single molecule containing all three rRNAs, a 5′ leader sequence, a 3′ trailer

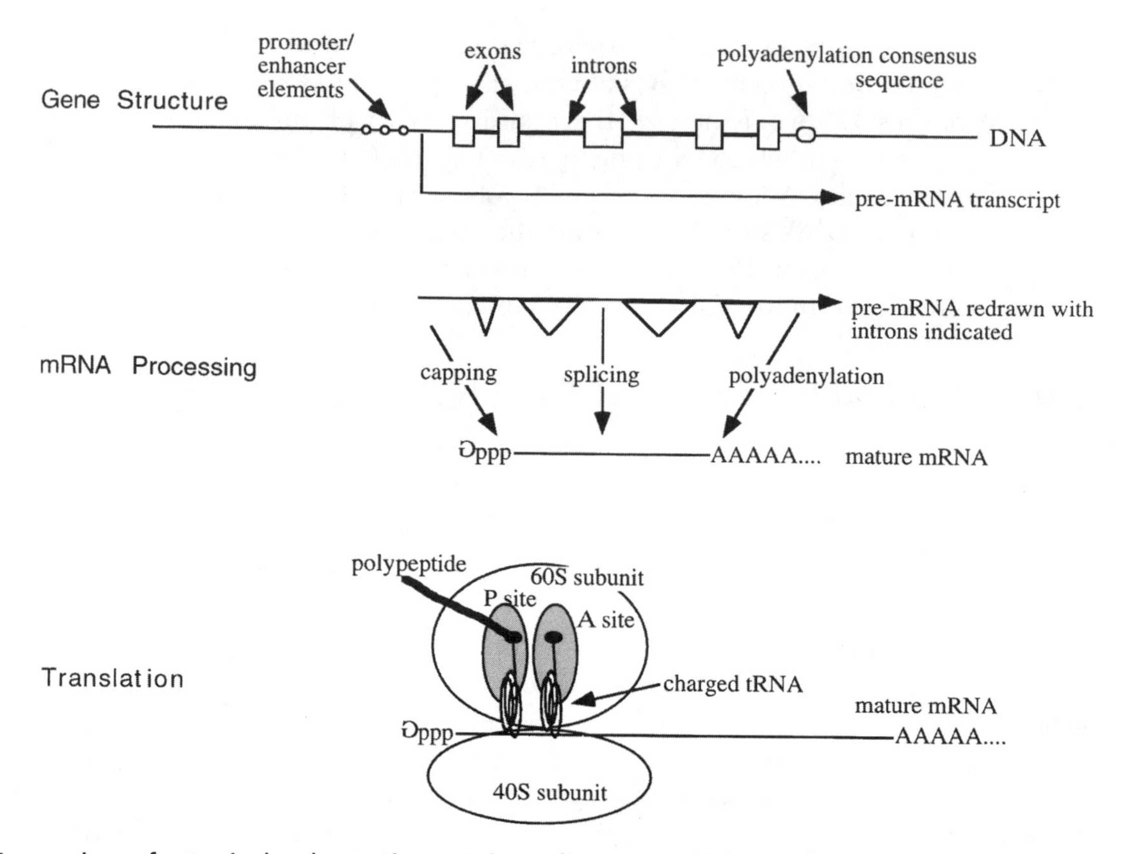

Figure 5.3 Expression of a typical eukaryotic protein-coding gene. Eukaryotic Gene Structure: Positions of promoter/enhancer elements are variable; one option is shown. Boxes represent exons. Heavy lines represent introns that will be removed from the mature mRNA. The position of the RNA transcription start site is indicated along with an arrow to show direction of transcription.
mRNA Processing: Pre-mRNA is redrawn to indicate splicing pattern. Triangles span the sequences that will not be contained in the mature mRNA. One processing event ("capping") adds a special guanosine nucleotide to the 5′ end of the transcript. Another processing event ("splicing") removes the introns. A third processing event ("polyadenylation") cleaves the pre-mRNA at the polyadenylation consensus sequence (downstream of the coding sequence) and adds a string of ~200 A nucleotides to the 3′ end of the transcript.

sequence, and internal spacers, some of which contain one or two **transfer RNA** (tRNA) genes. Organization of this sort ensures equal amounts of all three rRNAs. This precursor rRNA (pre-rRNA) is cleaved into the mature 16S, 23S, and 5S rRNAs and tRNAs by the enzyme **RNase III**. Mature transcripts that specify protein (**messenger RNAs** or **mRNAs**) consist of a leader sequence (5′, or **upstream**, of coding sequence), the coding region, and a trailer sequence (3′, or **downstream**, of the coding region). The transcript is not modified after transcription and it may contain the instructions for more than one protein (i.e., it may be **polycistronic**). Since transcription does not occur in a nucleus in prokaryotes, ribosomes can begin translating the message before it is completely synthesized!

Transcription is more complicated in eukaryotes. **Figure 5.3** shows one of the major kinds of transcription and may aid in reading the following. There are three types of RNA polymerase in eukaryotes, distinguished by the types of genes they transcribe. Each one is a complex of multiple subunits and does not itself bind to a gene's promoter. Members of a special class of proteins called **transcription factors**, bind to the promoter, and the polymerase binds to the transcription factors. Transcription factors therefore provide specificity to eukaryotic gene expression and are extremely important to the organism.

RNA polymerase I transcribes three of the four genes that encode rRNAs (28S, 18S, and 5.8S rRNAs). The rRNAs are integral parts of the ribosome, where protein synthesis occurs (see Topic 5). These rDNAs (i.e., the genes encoding rRNAs) are organized as tandem repeats in the order 18S, 5.8S, 28S, with **nontranscribed spacer** sequences between arrays and **transcribed spacer** sequences separating genes within arrays. The promoter and terminator for each transcription unit lies within the nontranscribed spacer. Each array acts as a single transcription unit, with one transcript containing one copy of each of the three genes. After transcription, the spacers are removed by specific ribonucleases, resulting in the formation of equal numbers of each mature rRNA. In some species, the spacer can catalyze its own removal without the aid of any proteins. These loci and RNA polymerase I comprise the nucleolus, within which ribosome subunits are assembled before they are exported to the cytoplasm.

In eukaryotes, genes that encode proteins (**structural genes**) are transcribed by the enzyme RNA polymerase II to give mRNA and some small nuclear RNAs (snRNAs). These latter RNAs are involved in mRNA splicing. Transcription by RNA polymerase II requires one or more promoters, defined as the nearest control element to the place where transcription begins. Examples of eukaryotic promoters are the TATA box, CAAT box, and GC box. All genes have at least one of these promoters. **Enhancers** are DNA sequences that can be bound by regulatory proteins to enhance (i.e., increase) transcription. Unlike promoters, enhancers can function at fairly large distances from the coding region.

Eukaryotic mRNAs are linear, diverse in size, and modified posttranscriptionally. (The generation and processing of a typical eukaryotic mRNA are shown schematically in Figure 5.3.) Before processing, the transcript is called a **pre-mRNA**. First, the 5′ end of the transcript will be **capped** by the covalent attachment of a special nucleotide, 7-methyl guanosine (i.e., guanosine with a methyl group on one of the base's carbons). This unusual nucleotide is attached to the first nucleotide of the transcript by a 5′-to-5′ linkage via three phosphate moieties. The cap is required for binding of the mature mRNA to the ribosome for protein synthesis (see Topic 5). Secondly, there are no termination sequences in the DNA for structural genes. Consequently the 3′ end of the transcript is generated by cleavage downstream of the **polyadenylation consensus sequence** (AAUAAA), followed by addition of approximately 200 adenosine nucleotides to the newly created 3′ end. The primary transcript may also contain **introns**, which are regions of noncoding sequence that interrupt the coding sequence (**exons**). Some genes in lower eukaryotes and most genes in higher eukaryotes have introns. Introns must be removed for the mRNA to function properly as the template for protein synthesis. Removal is performed in a series of events known as **mRNA splicing**. [The mRNAs in mitochondria, trypanosomes, and some other organisms are sometimes specially processed (**edited**) as well.] Until the time when all of these modifications have been made, the transcript continues to be known as a **precursor** (pre-mRNA). Once the processing is finished, the mature mRNA is exported out of the nucleus, to where it can direct the synthesis of proteins on ribosomes located in the cytoplasm.

In eukaryotes, RNA polymerase III transcribes three kinds of genes: the tRNA genes, the 5S rRNA genes, and the snRNAs that are not transcribed by RNA polymerase II. The 5S rRNA is another component of the ribosome. The tRNAs are the molecules that match the mRNA code to the proper amino acids at the ribosome (see Topic 5). For both the 5S rRNA and tRNA genes, the gene's promoter lies within the gene itself, in the **internal control region**, or **ICR**. This region usually contains two functional motifs for the promoter and transcription factors that are necessary for transcription of these genes. The functional motifs are specific for the type of gene in which they occur, that is, tRNA or rRNA. The primary transcript from the 5S rDNA gene is mature when made and requires no processing.

The tRNAs make up 10% to 15% of the total RNA in both prokaryotes and eukaryotes. They are made as precursors (**pre-tRNAs**) that have 5′ leader and 3′ trailer sequences. Both are removed by enzymes before the tRNA is mature. Some tRNA genes in some species have introns. These are removed by a pair of enzymes that are not involved in mRNA splicing. The mature tRNA is a single molecule, 75 to 90 nucleotides in length, whose nucleotides have been modified considerably. Two parts of a tRNA are notable: the anticodon loop and the 3′ end. The anticodon is contained in a looped (not base-paired) portion of the molecule. It consists of a three-nucleotide sequence that is complementary to and pairs with a three-nucleotide sequence in mRNA. This activity enables the tRNA to deliver the correct amino acid to the ribosome for incorporation into a nascent protein (see Topic 5). The second notable feature is located at the 3′ end of all tRNAs: the sequence 5′-CCA-3′. This is where an amino acid is temporarily covalently attached to the tRNA. Even though there could be as many as 61 different tRNAs in a cell (see Topic 5), extensive base pairing within the molecule causes all of them to assume a typical three-dimensional structure (like a capital L), that resembles a cloverleaf when written as a base-paired nucleotide sequence.

Topic Test 4: RNA and Transcription

True/False

1. Most eukaryotic transcripts have to be processed to be biologically active.
2. RNA polymerase III transcripts are *E. coli* mRNAs.

Multiple Choice

3. The following general features of RNA polymerases are true of all such enzymes, but one statement is true of prokaryotic but not eukaryotic polymerases. Which one?
   a. They do not require a primer.
   b. They bind a promoter sequence in the DNA.
   c. They are composed of multiple peptides.
   d. They transcribe one of two DNA strands.
   e. They synthesize in the 5′ to 3′ direction.
4. Which of the following is not a modification made to eukaryotic mRNAs?
   a. The 5′ leader sequence is removed.
   b. A cap is added to the 5′ end.
   c. Introns are removed.
   d. The 3′ end is cleaved at a specific site.
   e. Multiple adenosine nucleotides are added to the 3′ end.
5. Which molecules carry the information for proteins to be synthesized?
   a. mRNAs
   b. rRNAs

c. tRNAs
d. mRNAs and tRNAs
e. All of the above

Short Answer

6. What is a promoter? What is a terminator?
7. In what ways are tRNAs distinctive?

Topic Test 4: Answers

1. **True.** The mRNAs are capped, spliced, and polyadenylated. Three of the four rRNAs are clipped out of the large precursor that contains all three, separated by spacers. The tRNAs are trimmed of their leader and trailer sequences.
2. **False.** RNA polymerase III is the eukaryotic enzyme that transcribes tRNAs, some snRNAs, and the 5S rRNA.
3. **b.** Prokaryotic RNA polymerases bind the promoters of genes. Eukaryotic polymerases do not bind promoters directly; they bind the transcription factors that bind the promoters.
4. **a.** Eukaryotic mRNAs do not have 5′ leader sequences. The tRNAs do.
5. **a.** The mRNAs are made from protein-coding genes. The rRNAs and tRNAs are functional as RNAs.
6. Promoters are the sequences in DNA that signal the beginning of a transcription unit to RNA polymerases. In prokaryotes, these sequences interact with the polymerase. In eukaryotes, promoters are bound by transcription factors; the polymerase interacts with the transcription factors. Terminators are sequences in prokaryotes that are found at the 3′ ends of genes which cause the termination of transcription.
7. The tRNAs are distinguished from other RNAs in several ways. These RNAs are the products of RNA polymerase III in eukaryotes. The nucleotides that comprise tRNAs are frequently modified. Base pairing within the tRNA molecule creates a cloverleaf structure, which is folded into an L shape. One portion of the molecule has a three-nucleotide sequence (an anticodon) that base pairs with mRNA to deliver an amino acid covalently bonded to the 3′ end of the tRNA to the ribosome.

TOPIC 5: PROTEIN AND TRANSLATION

Key Points

✓ *What are proteins?*

✓ *How are proteins made?*

The second part of the central dogma concerns the conversion of the information in mRNA into protein. This is **translation**.

A protein is composed of one or more polypeptides. Polypeptides are built from amino acids by the translational machinery according to the instructions contained within mRNA. The mRNA

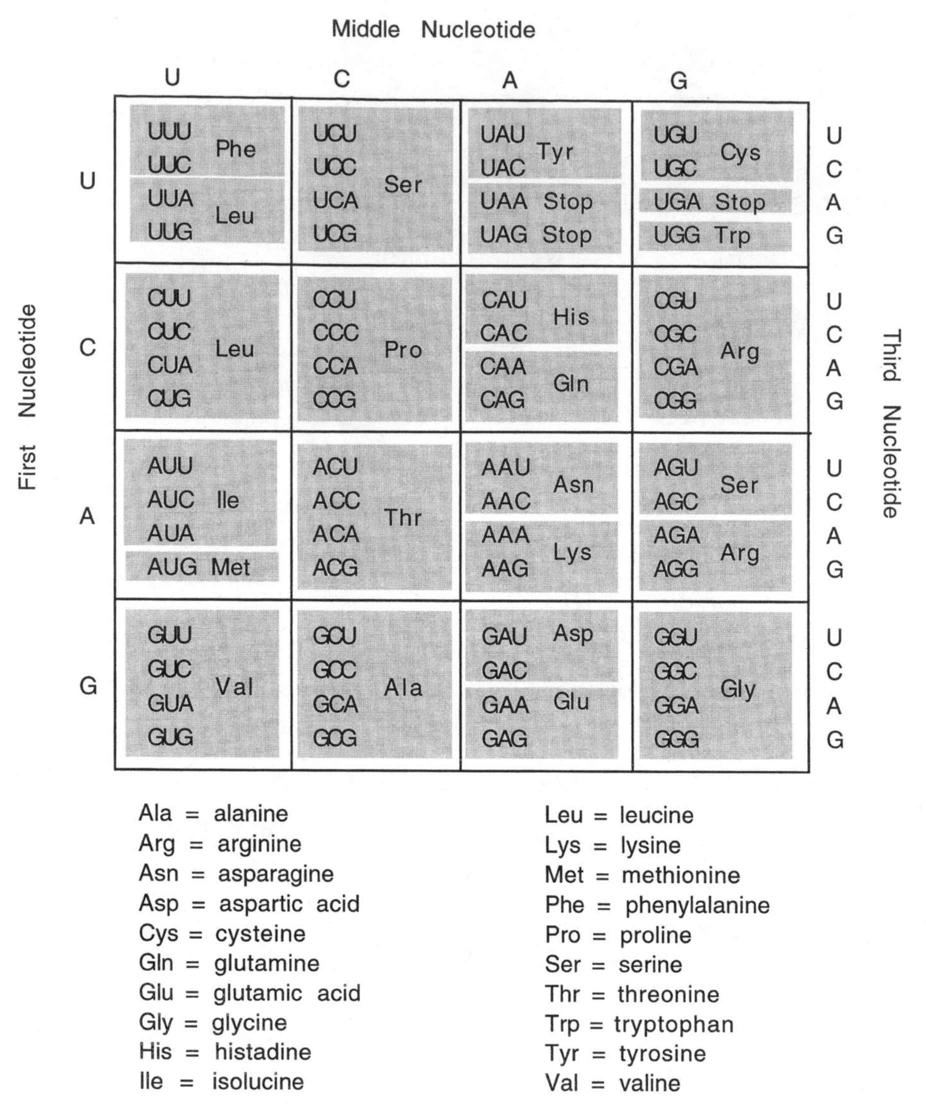

Figure 5.4 The genetic code.

sequence is determined by the DNA sequence, so the DNA sequence of a protein is **genetic code**. **Amino acids** are nitrogen-containing organic acids. Those that compose proteins are structurally related. The general formula for an amino acid is H_2N-CHR-COOH. **Proteins** are chains of amino acids in which one residue's carboxylic acid group is bonded to the next one's amino group. The middle carbon atom in an amino acid is bonded to a hydrogen atom (H) and to a variable group (R, the general symbol for any reactive group) that gives the particular amino acid its identity. Among the 20 amino acids commonly found in proteins, R can be as simple as the hydrogen atom in glycine, or as complex as the double-ring structure of tryptophan.

The genetic code is triplet, degenerate, and nonoverlapping. Each amino acid is encoded by three consecutive nucleotides (i.e., triplets). Since there are four nucleotides in DNA and RNA, there are 64 possible triplets, called **codons**. Only 20 amino acids arc commonly found in proteins, so there would seem to be surplus codons. Most amino acids are specified by more than one codon (**Figure 5.4**), meaning that the code is degenerate. Three of the codons do not specify any amino acid, and one specifies a special-use amino acid. AUG encodes methionine and is also the **initiation** (or **start**) **codon**; that is, it signals the beginning of the coded protein sequence. **Stop codons** signal the end of the polypeptide and are UAA, UAG, and UGA. These are also called **nonsense codons** and **chain-terminating codons**. There are no tRNAs that read stop codons. The remaining codons are **sense codons**; these specify amino acids. The code is nonoverlapping in that three adjacent nucleotides comprise one codon; that is, they determine the identity of each amino acid. The next three amino acids specify the next amino acid in the protein's sequence. Nucleotides are not skipped between codons, or reused (codons do not overlap each other). Because the code is read in triplets, reading frame is an important concept. The start codon determines the reading frame for an mRNA sequence, that is, which of the three possible sets of triplets specifies the protein. The stop codon must be in the same reading frame as the start codon. A general feature of coding sequences is that they only "read-through" in one of the three possible triplet codon frames. In the other two frames, there are numerous stop codons. This helps scientists to determine the protein sequence from a given DNA sequence when the reading frame is not known. The code is nearly universal in that almost all species use the same correspondence between codons and amino acids or chain terminators. Finally, there could be 61 tRNAs in a cell, one for each sense codon. In fact, there are fewer, because the tRNA-mRNA pairing is imperfect in the third position of the anticodon: The tRNA's anticodon is said to "wobble." For example, G in the tRNA is allowed to pair with C or U in the mRNA. Notice in the codon table (see Figure 5.4) that degenerate codons are most likely to vary only at the third position. The advantage to wobble is that, for example, the tRNA whose anticodon is 5′-GGC-3′ binds 5′-GCC-3′ in the mRNA, as well as 5′-GCU-3′, both of which specify alanine. A single tRNA translates both codons correctly.

Translation is better understood in prokaryotes than in eukaryotes, so this description focuses on prokaryotes. There are three major components to the translation machine: mRNA, tRNAs, and ribosomes. Prokaryotic mRNAs are not modified after transcription, and their translation can begin while their transcription is nearing completion because there is no nucleus (this is **coupled transcription-translation**). The protein sequence is coded in the mRNA linearly so that the amino terminus (N terminus) is near the 5′ end of the mRNA and the carboxyl terminus (C terminus) is near the 3′ end. The tRNAs are the adapters that match codons with appropriate amino acids. The enzyme **aminoacyl-tRNA synthetase** exists in 20 types, one for each amino acid. This enzyme attaches the carboxyl group of an amino acid to the 3′ end of the tRNA, creating a **charged tRNA**, or **aminoacyl-tRNA**.

The cellular apparatus that performs translation is the **ribosome**. (See Figure 5.3 for an illustration of a eukaryotic ribosome. Though the process is not identical, prokaryotic and eukaryotic ribosomes are similar enough that this figure will help with understanding translation in both types of organisms.) Ribosomes are complexes formed from multiple polypeptides and, in prokaryotes, three rRNAs. They exist as two subunits. The large subunit (50S) contains the 23S and 5S rRNAs and 34 polypeptides. The small subunit (30S) contains the 16S rRNA and 20 polypeptides. Both protein and RNA components of the ribosome are essential to its function. When the ribosome is not actively engaged in translation, the ribosome subunits are separated from each other.

In prokaryotes, the first step in initiation of translation is the binding of mRNA by the small subunit, to which three initiation factors and guanosine triphosphate (GTP) are attached. Upstream (5′) of the initiation codon, the mRNA contains a ribosome-binding site (the **Shine-Dalgarno sequence**) that base pairs with the 3′ end of the 16S rRNA. This sequence is required for binding of the mRNA and helps the ribosome to identify the nearby AUG initiation codon. It specifies a formylmethionine amino acid, which is usually the first amino acid in the chain. Formylmethionine is brought to the ribosome on a special tRNA that is only used for initiation; methionine codons within the protein are read by a separate tRNA that is charged with normal methionine. This special tRNA binds to the existing complex to form the 30S initiation complex, causing the release of one of the initiation factors. Then the large subunit binds to the complex, causing the release of the other two initiation factors, hydrolyzing the GTP, and forming the 70S initiation complex. Aminoacyl-tRNAs bind two sites on the 70S complex: The A site binds single *a*minoacyl-tRNAs and the P site binds the aminoacyl-tRNA that is attached to the growing chain of amino acids (i.e., the *p*eptide). (When the nascent peptide is attached to a tRNA in the P site, that tRNA is a **peptidyl-tRNA**.) At the time of initiation, the 70S complex has the first aminoacyl-tRNA (formylmethionine) in the P site and is competent for protein synthesis.

The elongation phase of protein synthesis consists of reiteration of three steps: binding of a charged tRNA, peptide bond formation, and **translocation** to the next codon. The appropriate aminoacyl-tRNA binds to the exposed codon in the A site by hydrolyzing one GTP and releasing the elongation factor EF-Tu, which accompanied the aminoacyl-tRNA to the ribosome. Peptide bond formation occurs in two steps. First, the bond between the carboxyl group of the amino acid in the P site and its tRNA is broken. Second, this carboxyl group is bonded to the amino group of the aminoacyl-tRNA in the A site. This **peptide bond** forms the backbone of the polypeptide, with the R groups extending out from it. Formation of the peptide bond requires the activity of peptidyl transferase; the 23S rRNA is part or all of this enzyme. The products of this reaction have differing fates during translocation. The newly "uncharged" tRNA will be released from the P site. The peptidyl-tRNA will be moved from the A site to the P site at the same time as the ribosome moves to the next codon in the 3′ direction of the mRNA, and a new aminoacyl-tRNA will bind the codon now occupying the A site. The process will repeat until the A site is occupied by a stop codon. A single mRNA molecule can be translated many times before it is degraded, and more than one ribosome can translate a single mRNA at one time. The average *E. coli* mRNA is used to synthesize about 10 proteins (from an equal number of ribosomes) simultaneously. Collectively, this complex of mRNA and multiple ribosomes is known as the **polysome**, or **polyribosome**.

Translation is terminated when the ribosome encounters a stop codon (UAA, UAG, or UGA), bound by one or more **release factors**, in the A site. (Remember, there are no tRNAs for these codons.) Release factors are proteins that cause the ribosome to release the polypeptide from the peptidyl-tRNA and to release the tRNA from the P site. Finally, they cause the ribosomal subunits to dissociate, releasing the mRNA. The released polypeptide is normally processed at least by removal of the initiating formylmethionine.

Translation in eukaryotes is consistent with what has been described for prokaryotes with the following exceptions. Eukaryotic mRNA is transcribed and processed (capped, spliced, and polyadenylated; see Topic 4) in the nucleus. It will be translated into a protein in the cell's cytoplasm after being transported out of the nucleus. Ribosomes in eukaryotes contain four rRNAs, organized as 60S and 40S subunits. The large subunit (60S) contains the 28S, 5S, and 5.8S rRNAs and approximately 50 polypeptides. The small subunit (40S) contains the 18S rRNA and approximately 35 polypeptides.

Initiation differs in eukaryotes in several ways. A **eukaryotic initiator factor (eIF)**, which contains several polypeptides including the **cap-binding protein**, binds to the cap (7-methyl guanosine) at the 5′ end of an mRNA. The small ribosomal subunit, many eIFs, GTP, and the initiator tRNA (charged with methionine, not formylmethionine) all bind to the mRNA and scan along it for the initiator AUG codon. When this codon is found, the large subunit binds the complex with concomitant release of eIFs. The mature initiation complex is 80S in size and has the first aminoacyl-tRNA, charged with methionine, bound in the P site. Eukaryotic elongation is similar to the prokaryotic mechanism, except that the peptidyl transferase contains the 28S rRNA, instead of the prokaryotic 23S RNA. After translational termination, processing of the released polypeptide can be very complex. Normally, the initiating methionine is removed. Processing may also involve cleavage into smaller fragments, or modification by phosphate or carbohydrate moieties, for example. Eukaryotic proteins may also be targeted for specific subcellular compartments by **signal sequences** or by **transit sequences** that are present in the amino acid sequence of the protein.

Proteins in both prokaryotes and eukaryotes can be comparatively tiny (e.g., met-enkephalin, produced by nerve cells, is 5 amino acids long in its active form) or huge (e.g., dystrophin, the muscle protein defective in muscular dystrophy, is 3,685 amino acids long). Their structures are described on four levels. The primary structure is the linear sequence of amino acids. The secondary structure is the spatial positions of neighboring amino acids. Two common secondary structural motifs in proteins are the alpha-helix and the beta sheet, both of which are stabilized by hydrogen bonding between carboxyl and amino groups within the protein. The tertiary structure is the three-dimensional shape of the entire polypeptide. If a protein interacts with other proteins to form a complex, it also has a quaternary structure. The quaternary structure is the shape of the complex formed when two or more polypeptides bind to each other to form a multimeric protein. The function of a protein is determined by its structure, and the structure is a function of its primary sequence, which in turn is determined by the sequence of the gene encoding the protein. Thus, protein functions are controlled by genes. (Experimental proof of this relationship is described in Chapter 8.)

Topic Test 5: Protein and Translation

True/False

1. Translation ceases at the chain-termination codon, corresponding to the amino terminus of the protein.
2. The amino acids of a protein are attached via their R groups.
3. Degeneracy means the genetic code is variable in different species.

Multiple Choice

4. One of the following is false. Translation
   a. initiation usually occurs at AUG.
   b. begins at the carboxyl terminus of the peptide and proceeds to the amino terminus.
   c. can occur multiple times from a single mRNA.
   d. is coupled with transcription in prokaryotes.
   e. initiation requires a Shine-Dalgarno sequence in the prokaryotic mRNA.
5. Which of the following is not a component of the translation apparatus?
   a. Aminoacyl-tRNAs
   b. Small ribosomal subunit

c. 5S rRNA
d. Initiation factors
e. snRNAs

Short Answer

6. What is the translation of the following mRNA sequence? Assume the first three nucleotides are the first codon.

   5′-AGU UUC GUU CGA AUG ACU GAA-3′

7. What is the significance of the AUG codon?

Topic Test 5: Answers

1. **False.** The chain-termination codon is where translation ends but this is the carboxyl terminus of the protein.
2. **False.** One amino acid's amino group bonds to the previous amino acid's carboxyl group.
3. **False.** Degeneracy means one amino acid may be encoded by more than one codon.
4. **b.** This statement is backward. The amino terminus is synthesized first and the carboxyl terminus is last.
5. **e.** The snRNAs are involved in splicing pre-mRNAs into mature messages in eukaryotes.
6. Ser-Phe-Val-Arg-Met-Thr-Glu.
7. This is the initiation codon, signifying the beginning of the peptide for most mRNAs in prokaryotes and eukaryotes. In prokaryotes, it is translated into formylmethionine by a special tRNA. In eukaryotes, it is translated into methionine.

TOPIC 6: CONTROL OF GENE EXPRESSION

KEY POINTS

✓ *How do organisms or cells express particular genes when they are needed?*

✓ *How do organisms or cells prevent particular genes from being expressed when they are not needed?*

Some general principles have emerged from studies of gene expression in bacteria and phages. First, related genes can be coordinately regulated. Secondly, gene expression can be controlled negatively through repression when the gene product is not needed. Thirdly, gene expression can be controlled positively by directly activating a gene's expression. Lastly, attenuators can enable graded expression of a gene by causing premature termination of transcription. All of these mechanisms (and more) exist because organisms have to be able to adapt to changes in their environment. Furthermore, differences in gene expression are responsible for differentiation of tissues in multicellular organisms.

Our example of both negative and positive transcriptional control is lactose metabolism in *E. coli*. Lactose is a sugar that *E. coli* will metabolize if glucose is unavailable. Normally the bacterium does not make the gene products (*lac* genes) that are needed for lactose metabolism.

When lactose is available and glucose is not, expression of these genes is induced. Induction is possible because there is a controlling site within the transcription unit that can be bound by a regulatory protein to prevent transcription (this is negative control). In turn, the regulatory protein is responsive to an **inducer** that will signal the need for those gene products. The *lac* genes are said to be inducible because the presence of the inducer relieves the transcriptional repression of the *lac* genes. In the case of lactose metabolism, there is a set of three adjacent genes that are only expressed if lactose is present and glucose is absent.

Lactose is a disaccharide, and one of the gene products, **β-galactosidase**, cleaves it into glucose and galactose. β-Galactosidase also isomerizes lactose into allolactose, which is the inducer in this system. A second gene product, **lactose permease**, transports lactose from the environment into the inside of the cell, where it will be metabolized by β-galactosidase. The third enzyme is **transacetylase** and is not important to this discussion. *E. coli* cells growing in the presence of glucose make almost no molecules of these proteins. If shifted to a medium containing lactose and no glucose, these cells will make several thousand molecules of each protein.

Non-wild-type (i.e., mutant) strains for these structural genes have defects in these functions. Most mutants for the β-galactosidase gene ($lacZ^-$) cannot convert lactose to galactose and glucose, but have unaffected permease and transacetylase. Most mutants for the permease gene ($lacY^-$) cannot transport lactose into the cell, but have unaffected β-galactosidase and transacetylase. Mutants for the transacetylase gene ($lacA^-$) are wild type for β-galactosidase and permease. Mapping of these genes determined the genes to be tightly linked and in the order *lacZ lacY lacA*. In fact, all three genes are transcribed as a single **polycistronic** mRNA instead of three mRNAs. Because this is a single transcription unit, there is only one promoter. The entire collection of genes, promoter, and operator (regulatory site) is an **operon**. Operons are coordinately regulated clusters of genes.

The most interesting mutant alleles affect expression of all three proteins. Some mutants express the genes constitutively; that is, they are expressed regardless of the presence of lactose. Other mutants express the genes poorly or not at all. The constitutively expressing mutations map to one of two regions: a small region upstream of the three protein-coding genes, called the ***lac* operator (*lacO*)**, or a gene-sized region farther away, called the **lactose repressor gene (*lacI*)**. Most of the that cannot induce any of the operon's genes map to the promoter for the genes, P_{lac}. Some other noninducible mutants ($lacI^s$ alleles) map to the lactose repressor gene.

Regulation of the operon occurs according to the following scenario. The *lac* repressor (protein) blocks the transcription of the operon (i.e., of *lacZ lacY lacA*) by binding to the *lac* operator (*lacO*). The operator is a DNA sequence between the transcriptional start site of the operon and the promoter. When this site is bound by the repressor, RNA polymerase can bind the promoter but cannot initiate transcription of the operon. Hence, mutations in either the repressor or the operator that prevent the repressor from binding to the operator will always allow transcription, resulting in constitutive expression. In the wild type, expression is regulated through interaction of the repressor with allolactose. The binding of allolactose by a molecule of the *lac* repressor induces an allosteric shift in the protein's conformation that prevents it from binding the operator. Consequently, the polymerase is able to initiate transcription. By turning off the repression, allolactose has induced transcription of the genes needed for lactose metabolism. This is negative control of gene expression because the default state is repression of these genes. The operon is negatively controlled by the repressor.

The interesting non-wild-type allele of the operator is $lacO^c$. This allele allows constitutive expression of the operon, because it is not capable of being bound by the *lac* repressor.

A special type of diploidy (**partial diploidy**) is possible in bacteria by the use of the **F′ plasmid**. The F′ plasmid is an extrachromosomal, self-replicating DNA molecule that can carry a limited number of genes. By placing a second operon copy or a second repressor allele on the F′ plasmid, researchers can test the effects of heterozygosity for any of the genetic components of this system. In the following genotypes, the plasmid alleles are given first and the chromosomal alleles are given second, separated by a slash, as is customary for diploid genotypes. The bacterial strain that is genotypically $F'\ lacP^+\ lacO^+\ lacZ^-\ lacY^+/lacP^+\ lacO^c\ lacZ^+\ lacY^-$ (where "+" denotes wild-type alleles, "–" denotes mutant alleles, and "c" denotes the constitutive operator allele) expresses β-galactosidase constitutively and permease inducibly. The β-galactosidase expression is constitutive because the chromosomal operon's $lacO^c$ allele cannot bind repressor, so the $lacZ^+$ and $lacY^-$ downstream of the *lacO* are expressed all the time. Since the permease allele is mutant, the enzyme's activity is not constitutive; it is inducible because the plasmid operon is controlled by a wild-type operator. The operator is said to be ***cis*-acting** because it only affects the operon to which it is physically attached. Key features of this relationship are (a) that the operator is a site near the protein-encoding genes, and (b) the operator controls only the genes that are nearby on the same molecule. Likewise, mutant promoters are also *cis*-acting, preventing transcription of the attached genes and not the genes carried on another DNA molecule.

Contrast this with what happens when the repressor is heterozygous. The bacterial strain whose genotype is $F'\ lacI^+\ lacP^+\ lacO^+\ lacZ^-\ lacY^+/lacI^-\ lacP^+\ lacO^+\ lacZ^+\ lacY^-$ expresses both β-galactosidase and permease inducibly. This is possible because the $lacI^+$ allele produces a protein, a diffusible product, that can repress both operons (as long as their operators are wild type). The repressor is ***trans*-acting** because it acts on an operator not physically attached (i.e., proximal) to it.

When lactose is absent, the *lac* repressor acts negatively on the *lac* operon to prevent its expression. But if lactose and glucose are both present, the *lac* operon still is not expressed. Expression occurs when repression is relieved *and* activation occurs. It is the absence of glucose that causes the activation of expression. Hence, as with many other regulated genes, this operon is subject to positive control. Activation cannot be accomplished with the *lac* repressor; it can only repress. A second protein must act as the activator. This protein needs to be responsive to a signal and needs to be able to bind to a specific site in the operon. For the *lac* operon, **catabolite activator protein** (CAP) is the protein component of the activator. CAP interacts with cyclic adenosine monophosphate (cAMP) in the following manner. If glucose is available, cAMP levels are low and most CAP is not bound to cAMP. If glucose is not available, cAMP levels are high and CAP binds cAMP. The CAP-cAMP complex is the activator. Transcription is activated when this complex binds a site in the operon upstream of the promoter. CAP's presence at this place on the DNA makes RNA polymerase bind better to the promoter, perhaps by bending the DNA, thus inducing transcription. Therefore, low glucose levels cause high CAP-cAMP levels, which activate transcription. This process was originally called **catabolite repression** because expression of the *lac* operon is directly responsive to the cell's metabolic needs: repressed when cAMP is low. Formally, this activation is positive control because an activating signal (the CAP-cAMP complex) causes expression.

The phenomenon of attenuation can be observed in the regulation of the repressible operon for tryptophan synthesis genes. This operon is repressible because it is not expressed when the amino acid tryptophan is available. Attenuation requires the nascent mRNA to be capable of forming secondary structures by base pairing within the leader sequence in the 5′ portion of the transcript. Depending on which of several possible structures form, RNA polymerase either terminates or continues. Attenuation is possible because transcription and translation are coupled in prokaryotes; it cannot occur in eukaryotes. The decision to terminate or to continue transcription is influenced by the progress of the ribosome that begins translating the mRNA shortly after RNA polymerase begins transcribing it. Following transcriptional initiation, RNA polymerase pauses in response to the first mRNA secondary structure. At this time, the ribosome loads onto the transcript and begins translation, and the polymerase resumes transcription. If tryptophan is limiting, there is insufficient charged tRNA and the ribosome stalls where it needs a tryptophan to continue. The polymerase proceeds, copying enough leader sequence to allow the second particular secondary structure to form that does not terminate the polymerase. If tryptophan is not limiting, the ribosome does not stall, so a third possible secondary structure forms, this time including the sequence know as the **attenuator**. This particular structure causes RNA polymerase to terminate transcription prematurely. The ribosome actually responds to levels of charged tRNAs but the effect is to allow transcription to continue and make the gene products or to prevent those gene products from being made via premature transcriptional termination. The ability to cause premature termination of transcription makes attenuators able to control the amount of expression of particular genes as a result of coupled transcription-translation.

The basic concepts of gene regulation through activation and repression are also true for eukaryotes. However, the details differ. Regarding gene organization, eukaryotes do not have operons. Secondly, there are many ways to regulate gene expression: transcriptional control, pre-mRNA processing, transport of the mRNA out of the nucleus, degradation (half-life) of the mRNA, translational control, posttranslational modification of proteins, and degradation (half-life) of proteins. Thirdly, eukaryotic genes have enhancers as well as promoters. Enhancers provide greater control of transcription, just as the CAP-binding site does for the *E. coli lac* operon. Both promoters and enhancers are bound by transcription factors that recruit RNA polymerase II. Finally, several aspects of chromatin structure are correlated with the transcriptional status of genes. Nucleosomes are disrupted at the promoters of actively transcribed genes. The promoter regions of actively transcribed genes are less likely to have methylated C nucleotides and are more sensitive to digestion by the enzyme DNase I (experimentally).

Topic Test 6: Control of Gene Expression

True/False

1. Activation is enabled by proteins that bind *cis*-acting elements to stimulate transcription.
2. Inducers relieve the repression of negatively regulated genes.
3. Regarding expression of the *lac* operon, *lacI* is *trans*-acting.

Multiple Choice

4. Which of the following means of regulating gene expression does *not* occur in prokaryotes?
   a. Repressor proteins repress expression of enzymes that are not needed.
   b. Activator proteins induce expression of enzymes that are needed.
   c. Translation controls transcription through attenuation.
   d. Genes in operons are coordinately controlled.
   e. The processing of mRNAs is controlled.

5. Deletion of the promoter of the *lac* operon in bacteria (with a single copy of the operon) would cause which of the following patterns of gene expression?
   a. β-Galactosidase and permease are never expressed.
   b. β-Galactosidase and permease are expressed inducibly.
   c. β-Galactosidase and permease are expressed constitutively.
   d. β-Galactosidase is not expressed and permease is expressed.
   e. β-Galactosidase is expressed and permease is not expressed.

Short Answer

6. Explain what happens at the *lac* operon when *E. coli* is grown in the presence of lactose and glucose.
7. What partial diploid genotype will produce β-galactosidase constitutively and permease inducibly?

Topic Test 6: Answers

1. **True.** Activation in prokaryotes requires an activator protein that can bind to a site near the promoter(s) of specific genes. Eukaryotic transcription factors are also activators of transcription.
2. **True.** In the *lac* operon example, the *lac* repressor binds to the inducer allolactose. This prevents the repressor from binding the *lac* operator, relieving repression.
3. **True.** The gene *lacI* encodes the *lac* repressor protein. Since the protein is the functional entity, it does not matter where its gene is located; thus, it is *trans*-acting.
4. **e.** Prokaryotes do not process mRNA transcripts (see Topic 5).
5. **a.** Deletion of the promoter will prevent RNA polymerase from binding so all expression of the operon is lost.
6. Lactose prevents the *lac* repressor from binding the operon. Glucose keeps the cAMP levels low so CAP is not complexed to cAMP. Since the CAP-binding site is not occupied, transcription is not stimulated.
7. $lacP^{+}\ lacO^{c}\ lacZ^{+}\ lacY^{-}/lacP^{+}\ lacO^{+}\ lacZ^{(+or-)}\ lacY^{+}$ and at least one $lacI^{+}$.

TOPIC 7: DNA REPAIR SYSTEMS

Key Points

✓ *What is DNA damage?*

✓ *How is DNA repaired?*

The genetic information is stored as a chemical (DNA or RNA) and, as with anything else, it is susceptible to damage. DNA polymerase may incorporate the wrong nucleotide, creating a mismatch in base pairing. Chemicals may damage the DNA, for example, by changing a base's chemical structure or covalently binding two bases together. Ionizing radiation such as x-rays may cause breaks in the DNA backbone or break a base away from its sugar. All of these damages to DNA need to be repaired to preserve the integrity of the genetic information.

When a nucleotide is damaged, it usually becomes incorrectly base paired. The alteration in base pairing or distortion of the double helix serves as the signal to the repair machinery that damage exists, and the correct nucleotide is recognized as such (see below). In this manner, the damage is specifically removed. If the damage is not corrected, it can be used as a template for DNA synthesis (see Topic 3), resulting in a permanent (heritable) change in the nucleotide sequence (an exception to this scenario is described later: postreplication repair).

Cells possess several enzymatic, genetically encoded systems for identifying and repairing damaged DNA. Five are described here. The first system is **photoreactivation**, which bacteria use to repair damage from ultraviolet (UV) light. One type of damage caused by the UV component of sunlight is the covalent joining (**cross-linking**) of adjacent pyrimidine bases (within one strand or across strands), making thymine dimers, cytosine dimers, and cytosine-thymine dimers. An enzyme in bacteria, DNA photolyase, uses the energy from visible light to remove the cross-links, restoring the bases to their normal bonding patterns.

All organisms have at least one type of the second mechanism: **excision repair**. Enzyme complexes recognize abnormal bases (chemically modified bases, thymine dimers, etc.) and cleave the DNA backbone of the damaged strand to remove the bad nucleotide(s). DNA polymerase fills in the roughly 10- to 30-bp gap left by the excision, and DNA ligase seals the remaining nick in the DNA backbone.

DNA polymerase can proofread its own work; if the polymerase detects a mispaired nucleotide in the strand it is synthesizing, it will usually remove the mismatched nucleotide before continuing. Unfortunately, DNA polymerase's proofreading is not perfect; it does not correct all of its errors. Hence, a third mechanism exists to recognize and repair these errors. This is the **mismatch repair system**. The enzyme complex that performs mismatch repair can distinguish the old strand from the new one via methylation. Most organisms methylate particular bases in their DNA; *E. coli* cells methylate the A in the sequence GATC, while mammals tend to methylate C in the sequence CG. Methylation of the newly synthesized strands is slow. Mismatch repair enzymes determine which strand is the original, by virtue of having more methylation than the newer strand, and then excise a chunk of the new strand including the mismatched nucleotide(s). As before, DNA polymerase fills in the gap and the nick is sealed by DNA ligase.

The fourth mechanism, the **postreplication repair system**, is a different sort of beast: It is a recombination system. If DNA polymerase is blocked during replication by DNA damage such as thymine dimers, the polymerase will abandon the nascent strand, and begin anew somewhat later in the template, skipping the damaged area. The recombination system recognizes this incompletely synthesized (**gapped**) region and recombines it with the duplex from the other side of the replication bubble. This recombination transfers a fragment of the correct, single-stranded region from the other duplex to fill the gap on the damaged duplex. The donor duplex is left with a gapped region, which is filled in by DNA polymerase. The transferred fragment can be used by the excision repair system as the template for repair of the original damage. If the original damage is not repaired before the next S phase, the postreplication repair system can repeat its performance.

Finally, we come to a mechanism that surprised geneticists: **error-prone repair**. There is a limit to the amount of damage that a cell can withstand in a defined period of time, even while trying to repair the damage by the methods just described. When the limit is surpassed, bacteria activate error-prone repair systems, such as the SOS response. This is an all-out effort to get through the present circumstance, so if replication is occurring, the cell tries to keep the DNA polymerase from stalling in the midst of the damaged template. Instead of pausing and skipping over the damage as usual, the polymerase copies it, making many more errors than usual, without the normal opportunities for repairing them. This increases the number of heritable errors that are made (see Chapter 6) but the individual cell may be more likely to live through the initial DNA damage than if it had not activated an error-prone repair pathway. In summary, diverse systems exist in organisms to repair the normal and extraordinary damage that can occur to DNA, and reduce the likelihood that the damage will be transmitted to descendants.

Topic Test 7: DNA Repair Systems

True/False

1. Each kind of DNA damage is repaired by a specific mechanism.
2. Damage to DNA must be repaired so that it will not become heritable.

Multiple Choice

3. Photoreactivation repairs
   a. gaps in the newly synthesized DNA strand.
   b. mismatches.
   c. pyrimidine dimers.
   d. DNA polymerase errors.
   e. massive damage.
4. The SOS response is an example of
   a. photoreactivation.
   b. excision repair.
   c. mismatch repair.
   d. postreplication repair.
   e. error-prone repair.

Short Answer

5. The human disease xeroderma pigmentosum (XP) is inherited as a recessive sensitivity to sunlight. On sun-exposed portions of their skin, affected individuals develop severe freckling that frequently progresses to cancer. What do you suppose could be the cause of XP?
6. Describe excision repair.

Topic Test 7: Answers

1. **False.** The mechanisms overlap in their activities. For example, pyrimidine dimers are repaired by both photoreactivation and excision repair.

2. **True.** DNA damage becomes heritable when it is separated from the correct strand by replication and DNA polymerase uses it as a template.
3. **c.** UV light induces adjacent pyrimidines to become covalently attached. The enzyme DNA photolyase uses energy from light to split the dimers apart.
4. **e.** The name *SOS response* refers to the emergency nature of large amounts of DNA damage. Only error-prone repair systems can tackle damage of this magnitude.
5. The features of XP are consistent with a defective damage repair system. Increased sensitivity to sunlight is consistent with an excision repair defect. In fact, XP patients seem to lack excision repair because they are also sensitive to x-rays, gamma rays, and DNA-damaging chemicals.
6. A small region of DNA containing the damaged base(s) is chewed out (leaving the other strand intact). DNA polymerase fills in the gap. DNA ligase seals the nick.

IN THE CLINIC

In the battle against infectious organisms, the best weapons are those that kill the infection-causing agent but do not harm the patient. This specificity can be achieved by targeting differences between the host's and the microbe's metabolisms. Bacterial infections are often treated with one or more of a large group of naturally occurring chemicals called **antibiotics**. One class of clinically important antibiotics acts by interfering with protein synthesis, specifically the prokaryotic, but not eukaryotic, ribosome. Streptomycin and gentamicin bind the small (30S) subunit and prevent initiation. Ribosomal subunits that are bound by streptomycin cannot bind the initiator tRNA, preventing further assembly of the complex. Streptomycin also affects elongation by causing inaccurate translation. Furthermore, already formed complexes (mature ribosomes) cannot dissociate as they normally do when reaching the end of the transcript. Instead, the ribosomes are trapped in a conformation that prevents further protein synthesis. Tetracyclines also inhibit protein synthesis, albeit in a different manner. These drugs block the A site on the ribosome, preventing binding by aminoacyl-tRNAs. This activity is not specific for prokaryotic ribosomes, but since bacteria actively transport these drugs into their cytoplasm (causing up to a 50-fold difference between the inside and outside of the cell), the dose that is required to kill the bacteria can be achieved at bloodstream concentrations that are not toxic to the host.

Chapter Test

True/False

1. Repression and activation of the lac operon directly affect RNA polymerase.
2. Wobble means there is more than one tRNA per codon.
3. A single type of RNA polymerase transcribes all prokaryotic genes.
4. All kinds of RNA are essential components of the translation process.

5. DNA replication is accurate because synthesis occurs in the 5′ to 3′ direction.
6. DNA is negatively charged.
7. The partial diploid whose genotype is F' *lacI*$^-$ *lacP*$^+$ *lacO*$^+$ *lacZ*$^+$ *lacY*$^-$/*lacI*$^+$ *lacP*$^-$ *lacO*^{c} *lacZ*$^+$ *lacY*$^-$ is inducible for permease expression.
8. Postreplication repair uses the methylation difference between old and new strands to correct mismatches.

Multiple Choice

9. Which one of the following statements is not true of most DNA?
   a. Pyrimidines pair with purines.
   b. Bases are in the center of the helix.
   c. The strands are antiparallel.
   d. The helix is stabilized by base pairing between strands.
   e. The backbone is an alternating chain of phosphate and ribose moieties.
10. Which one of the following DNA repair mechanisms does not involve DNA polymerase?
    a. Photoreactivation
    b. Mismatch repair
    c. Excision repair
    d. Postreplication repair
    e. Error-prone repair
11. Which protein is not required to synthesize the leading strand?
    a. DNA helicase
    b. SSB
    c. DNA primase
    d. DNA polymerase I
    e. DNA polymerase III
12. Which of the following statements is untrue?
    a. *E. coli* RNA polymerase completes transcription when it copies a terminator.
    b. Promoters for RNA polymerase III transcription units can lie within the coding sequence.
    c. Four rRNAs are made by RNA polymerase I as a single precursor transcript.
    d. Leader and trailer sequences must be removed from pre-tRNA transcripts.
    e. Eukaryotic mRNAs have a special nucleotide on their 5′ ends.
13. Which one of the following is not true? Negative control of gene expression
    a. allows expression of needed genes.
    b. prevents expression of unneeded genes.
    c. requires a repressor protein.
    d. is relieved by an inducer.
    e. requires secondary structure in the transcript.
14. Which of the following statements is untrue?
    a. Eukaryotic mRNAs are bound to the initiation complex by their caps.
    b. The carboxyl terminus of the peptidyl-tRNA is the next amino acid addition site.

c. The next amino acid to be added to the growing chain is in the P site of the ribosome.
d. The large ribosomal subunit binds to the smaller when the initiation codon is found.
e. Termination in eukaryotes requires a stop codon to be bound by release factor(s).

Short Answer

15. Describe initiation and termination of transcription in prokaryotes.
16. In what ways do RNA polymerase and DNA polymerase differ?
17. *E. coli* cells are exposed to strong UV light in the laboratory. After the exposure, half are placed in a dark incubator and half are placed in a light-filled incubator. Will their damage repair differ?

Essay

18. Compare and contrast DNA and RNA (mRNA, rRNA, and tRNA) molecules as they exist inside cells (consider structure).

Chapter Test Answers

1. **T** 2. **F** 3. **T** 4. **T** 5. **F** 6. **T** 7. **F** 8. **F** 9. **e** 10. **a** 11. **d** 12. **c**
13. **e** 14. **c**

15. RNA polymerase recognizes the prokaryotic promoter to initiate transcription. It begins an RNA chain near to and 3′ of the promoter. The chain is synthesized in a 5′ to 3′ direction. Termination occurs when the polymerase has copied an inverted repeat (the terminator), which base pairs with itself to form a hairpin structure in the RNA. Following this in the template is a string of T nucleotides. Terminators lacking this T string also require the termination protein rho.
16. DNA polymerase uses deoxyribonucleotides and thymidine to make DNA. It requires a primer to which it can add nucleotides. Its normal activity (replication) involves copying both strands of the DNA duplex. It proofreads and can correct its errors. RNA polymerase uses ribonucleotides and uracil to make RNA. It does not require a primer. Its activity is transcription and it only copies one DNA strand. It cannot proofread.
17. Yes, the *E. coli* cells will differ in the mechanisms they use to repair the damage. The cells left in the dark will only repair their pyrimidine dimers using the excision repair system. Those left in the light will use both the excision repair system and photoreactivation to repair the damage.
18. DNA and RNA are composed of nucleotides. In DNA, a nucleotide is a phosphate attached to the 5′ carbon atom of deoxyribose and a nitrogenous base (A, G, C, or T) attached to the 1′ carbon atom of deoxyribose. RNA nucleotides differ in two ways: The sugar is ribose (2′ hydroxyl instead of hydrogen in DNA) and the base U replaces T. The strand is formed by attachment of the phosphate moiety in one nucleotide to the 3′ carbon atom of the adjacent nucleotide. DNA is usually double stranded; most RNA is single stranded. Interstrand hydrogen bonding between nitrogenous bases causes DNA to

be double stranded. The paired bases form the interior of the double helix; alternating sugars and phosphates form the backbone. The two strands in a double helix are polar and oppositely oriented.

Inside cells the DNA is wound around histone octamers to form nucleosomes. The nucleosomes are organized into a 30-nm-diameter fiber whose exact structure is unknown. This fiber is packaged into a more complex structure to make interphase chromatin, the form of DNA that exists inside cells at all times except cell division. For cell division the chromatin is compacted still further to form cytologically distinct chromosomes.

RNAs exist complexed within and with ribosomes during translation. The tRNAs have complex secondary structures formed by hydrogen bonding within the molecule. The rRNAs are complexed with protein within ribosomes. The mRNAs are mostly single stranded, but they are double stranded in particular situations, as when they are involved in gene regulation.

Check Your Performance:

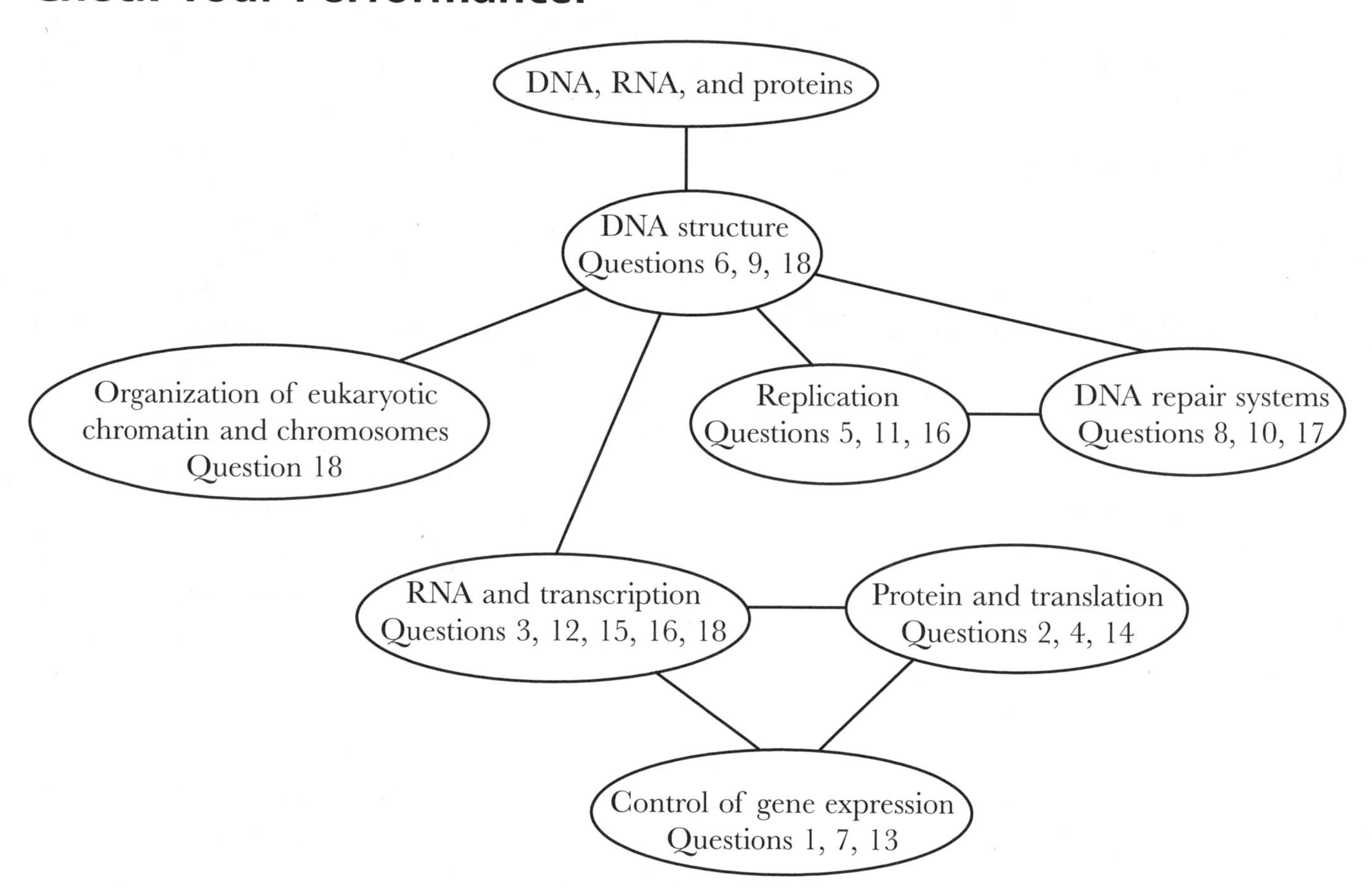

Use this chart to identify weak areas, based on the questions you answered incorrectly in the Chapter Test.

Mutation

Mutation is a physical or chemical change in a cell's DNA. At its most basic, a mutation is an alteration of the DNA sequence. It can be a small change, affecting one or many nucleotides of a single gene. This type of mutation is described in this chapter. Alternatively, mutations can be large, affecting several or many genes, or whole chromosomes. Large mutations are described in Chapter 7. This division of mutations into categories of one affected gene versus more than one affected gene is somewhat arbitrary. Both have important consequences for the organisms in which they occur. Mutation is a source of genetic variation and therefore phenotypic variation in species. Hence, it can provide the basis for the evolutionary divergence of species. On the other hand, mutations are a continuing source of ill health, by disrupting normal gene function. Fortunately, mutation is normally rare; the mutation frequency for a typical human gene (mutating to autosomal dominance) is about $1/10^6$ gametes. But this rate can be increased by **chemical mutagens**, which are substances that cause DNA damage, and hence can also be **carcinogens** (substances that cause cancer).

ESSENTIAL BACKGROUND

- **DNA repair (Chapter 5)**
- **Chemical structure of DNA (Chapter 5)**
- **Gene structure and translation (Chapter 5)**
- **Dominance and recessiveness (Chapters 1 and 3)**
- **Some organic chemistry nomenclature**

TOPIC 1: MUTATION

KEY POINTS

✓ *What are mutations?*

✓ *What are base-substitution mutations?*

✓ *What are frameshift mutations?*

Mutations are heritable changes in the nucleotide sequence of the genetic material of an organism, which is usually DNA. If the change occurs within the coding region of a gene, it can result in an altered amino acid sequence, that is, a new allele of the gene. Alternatively, mutations may be in the structural elements that control a gene's expression. Mutations range in size from those that alter a single nucleotide pair to those that affect whole chromosomes. This

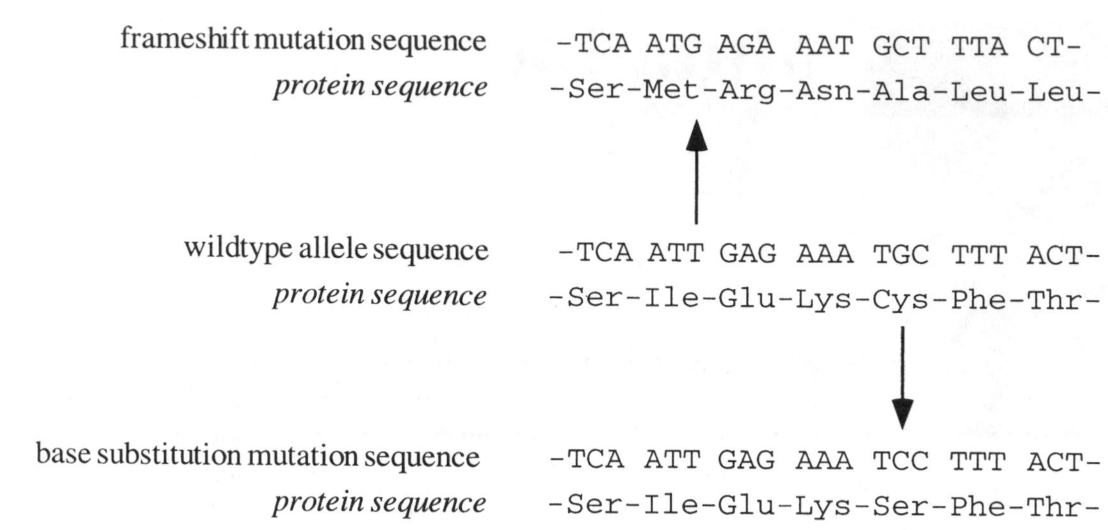

Figure 6.1 Two exampler of mutations. A portion of the wild-type sequence of a yeast gene occupies the center of this figure, along with its corresponding amino acid sequence. The nucleotide sequence is shown without the complementary strand and is divided into triplet codons solely for ease of illustration. The top sequence shows the effect of a frameshift mutation in which one of two consecutive T:A base pairs is deleted (the arrow indicates the location). Note the altered amino acid sequence to the right side of the mutation. The bottom sequence shows the effect of a G:C to C:G base-substitution mutation (the arrow indicates the location). This is both a transversion mutation, for the class of bases involved, and a missense mutation, because it changes a cysteine codon in the wild-type sequence into a serine codon in the mutant sequence.

chapter primarily concerns mutations that affect single genes, usually by altering just one or a few nucleotide pairs. Such **point mutations** are simple alterations in the nucleotide sequence.

Base-substitution mutations are the replacement of one nucleotide with another (see example in **Figure 6.1**). When this occurs within the coding region of a gene, it is called either a **missense** or a **nonsense** mutation. Missense mutations change the identity of an amino acid. Nonsense mutations create stop codons, causing the protein to be terminated at the point of the change. The resulting protein is shorter than normal and has lost part or all of its wild-type function. Genes are somewhat protected against the dangers of base-substitution mutations. In general, the more common an amino acid is, the greater the number of codons that encode it. Secondly, when there are multiple codons for an amino acid, they usually differ only at the third position, making mutations at this position less likely to alter the identity of the encoded amino acid. This reduces the likelihood of a random base change causing an amino acid change. Thirdly, amino acids that are chemically similar have similar codons. This means that if an amino acid is replaced by mutation, the new amino acid is likely to be chemically similar to the original, reducing the effect of the mutation. If a mutation does not change the amino acid sequence of the protein, due to codon degeneracy, we say the mutation is **silent**. If an amino acid is changed but the phenotype of the organism is wild type (i.e., gene function is not affected), we say the mutation is **neutral**.

The other kind of point mutation is the **frameshift mutation** (see example in Figure 6.1). Frameshift mutations are caused by the addition or loss of one or a few nucleotides within a coding sequence. If the number of nucleotides added or deleted is not a multiple of three, as it often is not, the frameshift mutation changes the reading frame of the gene. The reading frame is changed for all of the DNA that follows the mutation, resulting in altered amino acid coding.

The altered coding begins immediately adjacent to the mutation and continues until a new stop codon is reached, since the original stop codon is out of frame. The new termination (nonsense) codon can be earlier or later than the original stop codon. The garbled instructions after the site of the mutation plus the new nonsense codon mean that frameshift mutations are much more likely to cause serious problems in the encoded protein than are base-substitution mutations.

Most point mutations are fairly simple changes in the DNA sequence, so it is possible that the change can be reversed by a second mutation, which returns the gene to its original sequence (**true reversion**) or introduces a compensating mutation that restores protein function. This latter event is termed **intragenic suppression** because it suppresses the mutant phenotype rather than restoring the original DNA sequence. Note that reversion of frameshift mutations requires the addition of one nucleotide at or near the site of the deletion if the original mutation was the deletion of one nucleotide, and vice versa, so that the original reading frame is restored. Intragenic suppression of a frameshift mutation restores the wild-type phenotype by introducing another frameshift mutation at a different place in the gene to restore the original reading frame. The resulting protein sequence will be mutant between the two mutations but the protein sequence that precedes the first mutation and follows the second will be wild type. Base substitutions cannot revert frameshift mutations and frameshift mutations cannot revert base-substitution mutations.

Topic Test 1: Mutation

True/False

1. Mutation creates new alleles of genes.
2. Frameshift mutations change one amino acid.

Multiple Choice

3. Which one of the following features is not true of base-substitution mutations in coding regions?
   a. They can be reverted by another base-substitution mutation at the same site.
   b. They cannot be reverted by a frameshift mutation at the same site.
   c. They do not change the reading frame.
   d. They change one nucleotide to another.
   e. They change the amino acid encoded at that site and all of the ones that follow.
4. Which of the following will revert or suppress a frameshift mutation that is a one-base-pair addition to a gene's coding sequence?
   a. One-base-pair addition at the same site
   b. One-base-pair deletion at the same site
   c. Base-substitution mutation of nucleotide adjacent to the original site
   d. Base-substitution mutation anywhere
   e. More than one of the above
5. The wild-type sequence of part of a gene is ATCCTGATTCGAAGCCCCC. The sequence of a mutant allele in the same region is ATCCTGATTCGATGCCCCC. What kind of a mutation is this allele?

a. Frameshift mutation with one nucleotide lost
b. Frameshift mutation with two nucleotides added
c. Base substitution of A to T
d. Base substitution of A to G
e. Both a frameshift of one nucleotide lost and a base substitution of A to G

Short Answer

6. Differentiate between a missense and a nonsense mutation.

7. How are new stop codons created by base-substitution versus frameshift mutations?

Topic Test 1: Answers

1. **True.** Mutation changes the nucleotide sequence of a gene, thus defining a new allele. The allele may have the same phenotype as wild type because the mutation is silent or neutral, or it may have a different phenotype, as is usually the case for nonsense and frameshift mutations.

2. **False.** Frameshift mutations change the amino acid at the site of the mutation and all that follow.

3. **e.** This is a feature of frameshift mutations.

4. **b.** The deletion of the previously added base pair is a true reversion of the mutation. None of the other options will restore the correct reading frame.

5. **c.** At the thirteenth nucleotide, the wild type has an A nucleotide while the mutant has a T nucleotide.

6. A missense mutation changes the particular amino acid specified at the site of the mutation. A nonsense mutation creates a stop codon at the site of the mutation, causing termination of the polypeptide.

7. If a base-substitution mutation causes a stop codon, it alters a single nucleotide so that the codon containing that nucleotide is now a stop codon. Hence, the stop codon is at the site of the mutation. A frameshift mutation usually does not create a stop codon at the site of the mutation. Instead, it causes an out-of-frame stop codon that exists downstream of the mutation to become in-frame.

TOPIC 2: MUTAGENS

KEY POINTS

✓ *What are mutagens?*

✓ *How do mutagens cause mutation?*

✓ *How are mutagens identified?*

A **mutagen** is anything that increases the rate of mutation from the normal, low level of mutation. The normal, basal rate of mutation (number of mutations per unit of time) is low. One way that mutagens increase mutation rate is by damaging DNA. Specific mutagens tend to cause particular kinds of mutations. Chemicals and radiation are the examples of mutagens described below. The types of base-substitution mutations that these mutagens create can be classified as **transitions** and **transversions**. Transitions are the exchange of one purine for the other purine (e.g., A taking the place of G) or of one pyrimidine for the other pyrimidine (e.g., T for C). Transversions are when the new base is of a different chemical class than the old base (e.g., C replacing A).

Some common chemical mutagens are **alkylating agents** such as ethyl methanesulfonate (EMS). These chemicals covalently bind alkyl groups to the nitrogenous bases of nucleotides, thereby causing mispairing of the base. The consequence of mispairing is that incorrect nucleotides are incorporated during DNA synthesis, primarily resulting in specific base-substitution mutations (i.e., heritable changes). DNA repair systems within cells can repair much of the initial damage (see Chapter 5). DNA replication converts damage into mutation when it restores base pairing to the damaged region by changing the nucleotide opposite the damage (i.e., using a nucleotide other than the wild-type one).

Base analogs are another class of chemical mutagens. These chemicals are structurally similar to a normal base and are incorporated into DNA in place of the normal base. For example, 5-bromouracil can be incorporated into the DNA in place of thymine. It can base pair with adenine, as does thymine, but it frequently tautomerizes to the enol form, which pairs with guanine, causing transitions. If a 5-bromouracil moiety is repaired to cytosine because it is paired with guanine, a mutation is created.

Nitrous acid is an example of a third type of chemical mutagen, a **deaminating agent**. It deaminates adenine to hypoxanthine, which base pairs with cytosine, and deaminates cytosine to uracil, which base pairs with adenine. These mispairings can lead to transition mutations. Alkylating agents, base analogs, and nitrous acid are all examples of mutagenic agents that primarily, but not exclusively, cause base substitutions.

Acridine dyes are examples of mutagens that are more likely to cause frameshift mutations than base substitutions. These agents act by intercalating between the stacked bases and causing slippage in the pairing of nucleotides, especially during replication. This results in mispairing initially and mutation when the mispair is repaired incorrectly or replicated in error. Various forms of radiation are also mutagenic. Ultraviolet (UV) light causes adjacent thymine molecules to become covalently attached to each other, resulting in a **thymine dimer**. The thymine dimer is a precursor to the mutation; errors in the repair of the dimer, or problems replicating the DNA through it, cause the wrong nucleotides to be incorporated, resulting in mutation. X-rays and gamma rays, both forms of ionizing radiation, are well-known mutagens. They act by creating high concentrations of reactive ions along the paths they travel through tissues. These ions cause the initial damage to DNA, which can be either replicated or repaired incorrectly, creating mutations. Ionizing radiation can cause both point mutations and chromosomal mutations. Through the generation of reactive ions, this radiation breaks the bonds that hold DNA together, including breaking bases off sugars or breaking the DNA backbone. Repair of the breaks can create chromosomal rearrangements that affect multiple genes, such as those described in Chapter 7.

How can we identify mutagens? The most widely used test for detecting suspected mutagens was developed in the 1970s by Bruce Ames. The test uses disabled strains of the bacterium *Salmonella typhimurium*, which are unable to make their own histidine (they are *his*$^-$) and therefore cannot grow on a medium lacking this amino acid. Different strains have either a base substitution or a frameshift allele of a gene whose protein product is required for histidine synthesis. The allele causes the his^- phenotype. Exposing these bacteria to a mutagen will enable some of the bacteria to grow on the histidine-deficient medium by reverting their *his*$^-$ mutation. At low frequency, untreated cells will also revert; the test of the suspected mutagen is whether it causes *more* cells to revert than do so spontaneously. Each reverted cell will make a colony of its descendants, so the number of colonies that appear after the cells have experienced the chemical is a measure of the potency of the mutagen. Using different strains in the test enables detection of mutagens that cause either frameshift mutations, base-substitution mutations, or both. Finally, the livers of mammals are capable of making mutagens out of various substances that are harmless on their own. This is relevant because any substance that enters the human body may be metabolized in the liver. For this reason, substances used in the Ames test are frequently pretreated with a rat liver extract. The enzymes present in the extract can convert the substances into what they would become through the action of a body's own liver, thereby making the Ames test more reflective of the substances' mutagenicity in humans. This test of mutagenicity is powerful because it is rapid (about 1 week) and inexpensive (few dollars) compared to rodent studies, which can take 3 to 4 years and $1 to $2 million per substance tested. The Ames test is also accurate: More than 90% of mutagens in mammals (identified by rodent studies) test positive for mutagenicity in the Ames test.

Topic Test 2: Mutagens

True/False

1. Mutagens that cause frameshift mutations are detectable in the Ames test.
2. Acridine dyes primarily cause frameshift mutations.

Multiple Choice

3. Which one of the following mutagens does not act primarily through mispairing?
   a. Alkylating agents
   b. Base analogs
   c. Deaminating agents
   d. X-rays
4. Which of the following is a transversion?
   a. One-base-pair deletion
   b. A changed to G
   c. T changed to C
   d. C changed to G

Short Answer

5. Explain whether you think nitrous acid or an acrydine dye would cause more colonies to develop in an Ames test if the *Salmonella* utilized has a base-substitution mutation in a histidine-synthesis gene.
6. What are the advantages of the Ames test over rodent tests of mutagenicity?

Topic Test 2: Answers

1. **True.** Some *Salmonella* strains in the Ames test are frameshift mutants. Frameshifting mutagens can revert or suppress the original mutation.
2. **True.** These chemicals cause mutations when they slide in between the stacked bases of the double helix and cause slippage of the DNA strands. This slippage causes mispairing that is repaired or replicated into a mutation.
3. **d.** X-rays break the bonds that hold DNA together by ionizing the surrounding substances, which then attack the DNA. Mutation results from the cells' attempts to fix the damage.
4. **d.** Cytosine is a pyrimidine and guanine is a purine, so the base substitution of G for C is a transversion (a change in chemical class).
5. Nitrous acid would cause more *Salmonella* reverent colonies than would an acrydine dye. The base-substitution mutation in the *Salmonella* can be reverted more successfully by nitrous acid, which primarily causes base-substitution mutations. Acrydine dyes primarily cause frameshift mutations, which will not revert this *Salmonella* mutation.
6. The Ames test is *much* cheaper and faster than rodent tests, while still being accurate and allowing a way to take the effect of the liver into account.

TOPIC 3: FEATURES OF MUTATION

KEY POINTS

✓ *Can mutation be spontaneous?*

✓ *Is mutation always harmful?*

✓ *Is gene mutation a random process?*

✓ *What is the source of genetic variation?*

Several important points should be discussed. First, mutations can be spontaneous. The **basal mutation rate** reflects these mutations, and is expressed as the number of mutations per gene (or base pair) per generation (or other unit of time). Rates are organism and gene specific and typically range from 10^{-8} to 10^{-5} per cell division. Spontaneous mutations result from errors in replication and recombination or faulty repair of normal DNA damage. In addition to spontaneous mutation, mutation can also be induced by exposure to a mutagen (see Topic 2). Geneticists use potent chemical mutagens or radiation to create mutations for genetic studies. Among human populations, the existence of mutagens is suspected when a group of individuals has a mutation rate different from that of a similar group in different conditions (usually using cancer incidence as the measure of mutation). Yet, even if a mutagen is detected in one group's environment, it cannot be known whether any individual's specific mutation was caused by that mutagen. Mutations (alterations of the DNA sequence) are chemically no different whether spontaneous or induced in origin.

The second point is that not all mutations change the organism's phenotype, and even if they do, the change is not always for the worse. There are two reasons why a mutation might not have an effect in the organism in which it occurs: It might be neutral, or it might be recessive. If it is recessive, the phenotype will not be seen until an organism that has two copies of the mutant allele arises, through selfing or inbreeding. Even if the mutation does affect the phenotype, that phenotype can be either beneficial or deleterious and anything from mild to severe in effect.

The third point is that genes mutate randomly. Any gene, any part of any chromosome can mutate at any time. The extension of this point is that contrary to common belief, specific advantageous mutations do not arise in response to a stress, that is, as a way of adapting to that stress. Rather, those mutations that allow an organism to survive conditions that kill the wild-type individuals either preexist in the population or are the result of sudden increases in *overall* mutation rates in response to the stress. In fact, the randomness of gene mutation is an aid to geneticists. Mutation is used as a tool for studying biological processes through genetic dissection.

The fourth point concerns somatic and germline mutations. Of course, somatic mutations can be detrimental to the organism; prime examples are the mutations that lead to cancer. But somatic mutations cannot be inherited because somatic cells do not make gametes. Mutations in germ cells can be inherited, for good or bad for the species, and are useful to geneticists precisely because they are inherited. Studies of the effects of mutations reveal how organisms normally function. Finally, mutation is the source of the genetic diversity contained within all of the organisms in existence. Mutation creates the new alleles that alone, in combination with each other, and in combination with the environment give new phenotypes that can be retained or lost from populations through the process of natural selection (see Chapter 10).

Topic Test 3: Features of Mutation

True/False

1. Mutation occurs when organisms need specific new phenotypes.
2. Mutations occurring in somatic cells cannot be inherited.
3. Mutations are tools that geneticists use to study biological processes.

Multiple Choice

4. Most mutations that produce a different phenotype are
   a. beneficial.
   b. benign.
   c. harmful.
   d. equally likely to be a, b, and c.
   e. equally likely to be a and c.
5. Mutation is important for
   a. creating new alleles.
   b. deactivating mutagens.
   c. improving proteins.
   d. detecting *Salmonella.*
   e. all of the above.

Short Answer

6. How do somatic cell mutations differ from germ cell mutations?

Topic Test 3: Answers

1. **False.** Mutations occur randomly in DNA at any time; they are not produced specifically in response to a need.
2. **True.** Somatic cells do not form gametes and so cannot pass on their mutations. Only mutations in germ cells will be inherited.
3. **True.** By studying the way in which biological processes go wrong in mutants, geneticists learn what is required for the processes to work correctly.
4. **c.** The majority of mutations that alter the phenotype from wild type are harmful but are difficult to detect outside of the laboratory because they are also likely to be recessive.
5. **a.** Mutation is the source of new alleles in all species. The genetic diversity of a population results from mutation creating alleles and natural selection acting on them (described in more detail in Chapter 10).
6. Mutations that occur in somatic cells will not be inherited because somatic cells do not form gametes. Only germ cells form gametes, so only mutations that exist in these cells will be inherited.

TOPIC 4: CLASSIFICATION OF MUTATIONS

KEY POINTS

✓ *How are mutations described?*

✓ *What are the advantages of conditional mutations?*

✓ *Is one description true of all alleles of a gene?*

Geneticists classify mutations in a variety of ways. These classifications are descriptive, and most mutations can be described in several ways. For example, a single mutation can be forward, lethal, recessive, temperature sensitive, and null. In order to classify mutations, we need a point of reference. Wild type serves this purpose.

Mutations can be forward or backward. **Forward mutations** convert wild-type alleles to mutant alleles, and **back mutations** (or **reversions** or **backward mutations**) convert mutant alleles into wild-type alleles.

A special class of mutation is the **conditional mutation**. Conditional mutations (often, conditional lethal) only show the mutant phenotype when a particular condition exists. The most famous of these is the **temperature-sensitive mutation**, abbreviated **ts**. Many genes can be mutated to give ts alleles. This sort of allele gives a wild-type phenotype at one temperature (the **permissive** temperature) and a mutant phenotype at another, **restrictive** temperature. The phenotype of Siamese cats is produced by a ts mutation in a gene involved in pigment synthesis. These cats have dark paws, ears, noses, and tails because these areas of their bodies are sufficiently cool and thus the mutant gene product is active in pigment production. The pigment-producing protein is compromised in the warmer areas of their bodies, causing these areas to be lighter colored. Essential genes are difficult to study because the homozygous mutant phenotype is death. Conditional lethal alleles allow the homozygous mutant strains to be maintained at permissive conditions. When desired, the lethal phenotype can be studied by shifting individuals to the restrictive condition. Ts mutations are often base substitutions that cause just enough disruption in the protein's structure to make it active (usually) at lower temperatures but inactive at higher temperatures. A small fraction of ts mutations are active at higher temperatures but inactive at lower temperatures. These are called **cold-sensitive alleles**.

Another important class of conditional mutants are the **auxotrophs**. An organism is said to be a **prototroph** if it can survive on a minimal set of basic nutrients, making the complex molecules it needs from simple molecules. Most microorganisms are prototrophs. Auxotrophs cannot survive on simple molecules; they must have one or more complex molecules available in their environments, because they are mutant for one or more genes that are needed to synthesize these complex molecules. Our understanding of metabolism was greatly increased by studies of auxotrophic mutants of bacteria and fungi. These are also classified as **biochemical mutations**.

The vast majority of mutations that have phenotypes are recessive because it is generally easier to lose function than to gain it. Likewise, most mutations are also deleterious because it is easier to break something than to improve it. Mutant proteins that have less activity than the wild-type protein result from **loss-of-function mutations** and are generally recessive because one wild-type allele is usually sufficient for a wild-type phenotype. If the mutant has lost all function, the allele is said to be an **amorph**, or a **null**. A rarer kind of mutation gives a more active gene product. These alleles are called **gain-of-function** and are generally dominant because the new activity is present in the heterozygote, in excess of the wild-type activity. There are two kinds of gain-of-function alleles. One is an allele that is more active than wild type. The other kind of allele has a new activity, one not previously exhibited by the wild-type gene product.

A last point on this subject is that different mutant alleles of the same gene may fall into different categories according to this classification scheme. A diverse collection of alleles for a single gene is often beneficial for studies of gene function; if the alleles have distinguishably different phenotypes, they collectively indicate more about the normal gene's function.

Topic Test 4: Classification of Mutations

True/False

1. A mutation that converts a wild-type allele into a mutant allele is a forward mutation.
2. The permissive temperature is the condition at which a temperature-sensitive mutant strain is wild type in phenotype.

Multiple Choice

3. Which one of the following is not true? Temperature-sensitive mutant alleles
   a. express the mutant phenotype at one temperature.
   b. express the wild-type phenotype at a different temperature.
   c. can be made in most genes.
   d. may be cold sensitive.
   e. cannot confer a lethal phenotype.
4. Gain-of-function mutations are usually
   a. reversions.

b. forward mutations.
c. dominant.
d. permissive.
e. induced.

Short Answer

5. Describe the difference in terms of activity and dominance between loss-of-function and gain-of-function alleles.

Topic Test 4: Answers

1. **True.** Forward mutations are in the direction from wild type to mutant. Back mutation, or reversion, is a change from mutant to wild type.
2. **True.** *Permissive temperature* describes the condition that permits the wild-type phenotype in temperature-sensitive (conditional) mutants. The mutant phenotype occurs at the *restrictive temperature.*
3. **e.** One of the strengths of conditional alleles is the ability to maintain homozygous mutant lines at the permissive temperature, when the mutant phenotype is lethality.
4. **c.** These mutations are usually dominant because they exceed the activity of the wild type in a heterozygote.
5. Loss-of-function alleles produce gene products that have less activity (or no activity) than the wild type. Most of these alleles will be recessive because in the heterozygote, there will still be one wild-type allele producing the wild-type activity. Gain-of-function alleles produce gene products that have greater activity than, or a different activity from, the wild type. Most of these alleles will be dominant because "more activity" and new activities cannot be masked by wild-type activity.

IN THE CLINIC

Mutagens have proved to be effective in the treatment of certain diseases. The first anticancer drug to be used in patients was mechlorethamine, a nitrogen mustard. This and a number of other drugs used against cancer are DNA alkylating agents; their primary effect is to increase the rate of base-substitution mutations. One reason these drugs are effective is that cancerous cells grow much more rapidly than normal cells. More frequent DNA replication means there is a greater chance that DNA damage is not repaired before the DNA repair machinery can restore the original sequence. Eventually, each cancer cell accrues lethal mutations, and the disease goes into remission. Unfortunately, there is some evidence that normal cells also endure more mutation as a result of this cancer therapy. However, the choice of whether to be treated is an easy one, since the mutation risk to normal cells is much less significant than the mortality threat from the cancer.

Chapter Test

True/False

1. Point mutation is the alteration of a single site in a gene.
2. Ionizing radiation is mutagenic.
3. Induced mutations are physically unlike those that occur spontaneously.
4. Mutation rates vary by organism and by gene.
5. Mutations in germ cells will be inherited.
6. Some mutations are innocuous or beneficial.
7. The rate of mutation is increased by mutagens.
8. A conditional mutation is sensitive to the environment.

Multiple Choice

9. A mutant allele of a known gene encodes an abnormally short protein. What kind of mutation is most likely to be present in this allele?
   a. Missense
   b. Nonsense
   c. Neutral
   d. Reversion
   e. Somatic
10. Intercalating agents are most likely to cause what kind of mutations?
   a. Missense
   b. Nonsense
   c. Frameshift
   d. Reversion
   e. Somatic
11. To be inherited, a new mutation must
   a. occur in a germ cell.
   b. increase genetic diversity.
   c. be induced.
   d. be random.
   e. be recessive.
12. What kind of mutation is the replacement of a G:C base pair with an A:T base pair?
   a. Spontaneous mutation
   b. Nonsense mutation
   c. Frameshift mutation
   d. Transversion mutation
   e. Transition mutation
13. Which of the following is incorrect? Ionizing radiation
   a. is mutagenic.
   b. indirectly breaks bonds in the DNA.

c. forms reactive ions.
d. intercalates between the DNA base pairs.
e. includes x-rays.

14. Base-substitution mutations in coding regions that do not change the amino acid sequence are
a. neutral mutations.
b. silent mutations.
c. frameshift mutations.
d. missense mutations.
e. nonsense mutations.

15. Reversion of a frameshift mutation can be induced with which of the following substances?
a. Alkylating agents
b. Base analogs
c. Acridine dyes
d. Deaminating agents
e. Ionizing radiation

Short Answer

16. Scientists in a pharmaceutical company synthesize a novel chemical and want to determine its safety. What is the best method for initial testing of the substance's capacity for causing mutation?

17. How does true reversion differ from suppression?

Essay

18. Mutations in the *ade2* gene in the yeast *Saccharomyces cerevisiae* cause a red phenotype; wild-type yeast is white. [Yeast makes colonies like bacteria do (see Topic 2).] When single colonies of an *ade2* mutant strain are allowed to form colonies, two types of rare colonies appear among the large numbers of red colonies. One type of colony is completely white and the other type is sectored: mostly red with pie-shaped wedges of white. Explain these rare colonies.

19. What are some types of conditional mutations?

Chapter Test Answers

1. **T** 2. **T** 3. **F** 4. **T** 5. **T** 6. **T** 7. **T** 8. **T** 9. **b** 10. **c** 11. **a** 12. **e**

13. **d** 14. **b** 15. **c**

16. The Ames test should be one of the first tests they perform. If the substance is mutagenic, it is likely to increase the reversion frequency of histidine auxotrophy (*his*$^-$) in *Salmonella typhimurium* relative to the spontaneous frequency.

17. True reversion restores the wild-type DNA sequence. Suppression restores the wild-type phenotype by introducing a second, compensating mutation.

18. The completely white colonies are formed when a single reverted (*ADE2*, the wild-type allele) cell forms the colony. These are rare because reversion is infrequent. The sectored colonies are formed when a single descendant of the colony-founder cell reverts. This, too, is a rare event. All of the reverted cell's descendants are white and as long as none of the other cells in the colony revert, the descendants will form the only white sector in an otherwise red colony.
19. Cold-sensitive mutations are conditional mutations for whom the mutant phenotype is expressed at low temperatures. High temperatures cause the mutant organism to have a wild-type phenotype. Temperature-sensitive mutations are conditional mutations for whom the mutant phenotype is expressed at high temperatures. Cooler temperatures cause the mutant organism to have a wild-type phenotype. Other conditional mutations might be sensitive to oxygen concentration, or nutritional state, or some other facet of the environment.

Check Your Performance:

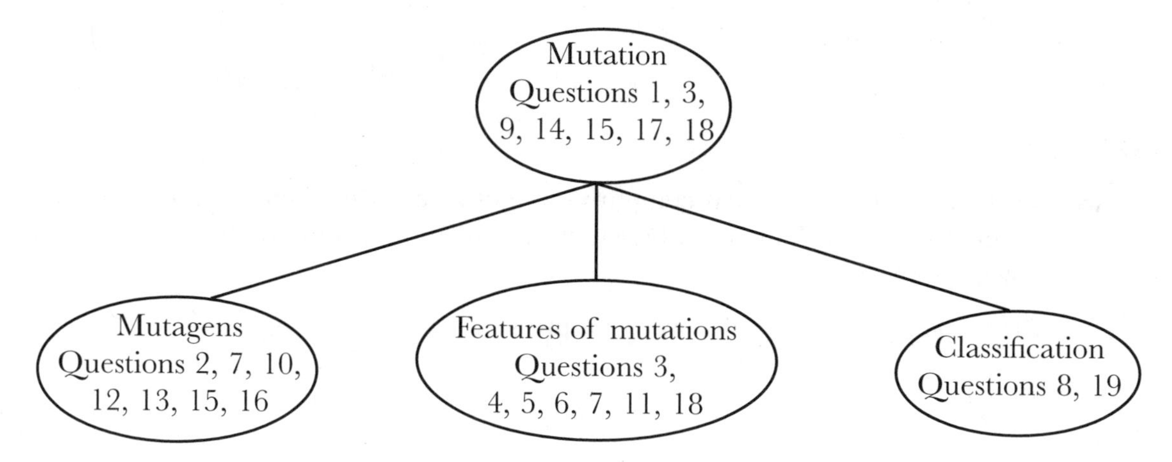

Use this chart to identify weak areas, based on the questions you answered incorrectly in the Chapter Test.

Chromosomal Aberrations

In Chapter 6, we saw the effect of mutation on single genes. These usually cause straightforward phenotypes. This chapter concerns mutations that occur on a larger scale. Some are even visible by microscopic analysis of stained mitotic chromosomes. Each of these mutations affects many genes, and therefore the resulting phenotypes are complex. These mutations are errors in chromosome number and errors in chromosome structure. Errors in number fall into two classes: mutations in the number of homologs or in the number of sets of homologs. Both of these are recognized by altered karyotypes. (A **karyotype** is the complete sets of chromosomes for a normal organism.) Errors in structure are classified into four types: deletion, duplication, inversion, and translocation. This chapter only addresses the alterations in chromosome number or structure that do not occur normally in wild-type individuals. They arise spontaneously and their frequencies can be increased by exposure to ionizing radiation (e.g., x-rays), and to a lesser extent nonionizing radiation and chemical mutagens. The complex phenotypes result from imbalances in gene dosage. Yet, in the whole of biology, some examples of these phenomena are not mutation, but rather are developmentally programmed modifications of the genome.

ESSENTIAL BACKGROUND

- Meiosis (Chapter 1)
- Linkage (Chapter 4)
- DNA structure (Chapter 5)
- Mutations and mutagens (Chapter 6)

TOPIC 1: POLYPLOIDY

KEY POINTS

✓ *What are polyploids?*

✓ *How do polyploids arise?*

✓ *What are the benefits of polyploidy?*

Ploidy is the number of sets of chromosomes possessed by an organism. One set is represented by n. If the organism has complete sets of chromosomes, it is said to be **euploid**. You already know the terms **haploid** (n, meaning one set of chromosomes, as in bacteria, some fungi, and gametes from diploids) and **diploid** ($2n$, meaning two sets of chromosomes, as in somatic cells in us and most other eukaryotes). A polyploid is an organism or cell that has more sets of chromosomes than is normal. **Triploids** ($3n$) have three sets of chromosomes. Triploids rarely make viable gametes. Random division of each trio of homologs results in one random cell having two homologs and the other cell having one. Hence, the usual result of segregation in a triploid is gametes with genetic imbalances that make them incapable of forming viable (living) zygotes upon fertilization. The probability of generating "good" gametes out of any meiosis in a triploid is the product of the probabilities of all of the other trivalents segregating in the same pattern as the first. Therefore, for an organism like the *Drosophila*, which has four chromosomes per haploid set, the probability of a triploid making a good gamete is $1/2 \times 1/2 \times 1/2$, or $1/8$. Most eukaryotes have more chromosomes, so their probabilities of good gametes would be much lower.

Tetraploids ($4n$) undergo meiosis much more successfully than do triploids, solely because they have an even number of chromosome sets. In fact, they are almost as successful at meiosis as are diploids. Ploidies higher than $4n$ are possible in plants, where polyploidy has been used to generate more desirable (i.e., larger) cultivars of food and ornamental crops.

Polyploids can arise through accidents of meiosis or can be made experimentally. Drugs, like colchicine, that stall meiosis for karyotyping by depolymerizing spindle fibers can be used to treat organisms briefly. When the treated cells are allowed to continue growing, they divide their nuclei without segregating the chromosomes properly. Many will be dead or aneuploid (see Topic 2) but some will be polyploid. Sometimes by chance one daughter cell will receive two copies of each chromosome. When that happens and that daughter cell combines with a normal gamete from the other parent, the progeny will be triploid. For example, a diploid parent whose germ cells are treated with colchicine will make some diploid gametes instead of haploid ones. These unusual gametes are fertilized by normal haploid gametes to create triploid zygotes. Fertilization by diploid gametes would produce tetraploid zygotes.

Tetraploids ($4n$) can be generated in either of two ways. The first is by doubling the number of chromosomes in a diploid ($2n$), resulting in a $4n$ organism. This can occur naturally (rarely) or through the use of colchicine. Tetraploids that arise in this manner are called **autotetraploid** (for other ploidies, **autopolyploid**). Another kind of tetraploid is known as an **allopolyploid** or **amphidiploid** and results from the doubling of an interspecies hybrid (and is therefore $2n_1 + 2n_2$). Increasing ploidy of gametes also solves another problem: Crosses between two species often result in infertile hybrids. Consider the mule: The mating of a female horse to a male donkey produces an animal called a mule that is sterile. The reason for the sterility in such hybrids is that chromosomes from the two parents are too dissimilar to pair during meiosis. Causing the ploidy to increase by two (i.e., making the diploid into a tetraploid) restores fertility by giving each chromosome a homolog with which to pair for meiosis. The products of this chromosome doubling have twice as many chromosomes as the normal pure species and are called amphidiploids. Because they now have two homologs of each chromosome, meiosis is successful and fertility is restored. This technique has enabled the amphidiploid grain *Triticale* to become an important crop. Take note, however, that nearly all polyploids are plants. Chromosome doubling will not help the mule to become fertile because this animal shares a limitation possessed by most higher-order animals: Polyploids are inviable. Polyploidy is not tolerated by mammals, birds, and most fishes, but it is common in amphibians, reptiles, and flatworms. Plants are famous for their ability to tolerate polyploidy; common examples include potatoes, strawberries, seedless watermelon, and octaploid ($8n$) giant chrysanthemums. The advantages to growers are that polyploids are larger than their diploid relatives and, if triploid, are also sterile (seedless). For example, wild bananas have large seeds that interfere with easy consumption. Commercially grown triploid bananas are infertile and therefore do not have unappetizing seeds. Wild oysters are unpalatable while they are releasing gametes, but sterile triploids do not go through this stage.

Topic Test 1: Polyploidy

True/False

1. All polyploids are sterile.
2. If an organism is autopolyploid, its polyploidy was induced (not spontaneous).

Multiple Choice

3. Which of the following statements is not true?
   a. The polyploid is larger than the diploid.
   b. Triploids are sterile.
   c. Tetraploids are not sterile.
   d. Polyploidy is common in plants.
   e. Polyploidy is common in mammals.
4. An organism that has two sets of chromosomes from one species and two sets from a second, related species is called
   a. diploid.
   b. triploid.
   c. autopolyploid.
   d. allopolyploid.
   e. none of the above.

Short Answer

5. What fraction of the gametes produced by a triploid can form normal (euploid) offspring, if the haploid number of chromosomes is two?
6. Hybrids formed by the mating of related species often are not fertile. Why not? What could make these hybrids fertile?

Topic Test 1: Answers

1. **False.** Some polyploids, like triploids, are sterile but those that have even numbers of chromosome sets are usually fertile.
2. **False.** An autopolyploid organism is one whose chromosomes come from a single species. The polyploidy could have occurred naturally or been induced.
3. **e.** Polyploidy is lethal to mammals.
4. **d.** This is an allopolyploid because the chromosomes originally came from more than one species.

5. One-half of the gametes can be fertilized to give offspring with complete sets of chromosomes. Four types of gametes are formed of genotype *AABB*, *AB*, *AAB*, and *ABB*. Euploid progeny could result from *AABB* and *AB*.
6. Interspecies hybrids have one complete set of chromosomes from each parent and these sets are sufficiently dissimilar that they cannot pair with each other for meiotic segregation. The haphazard segregation of the unpaired chromosomes causes the infertility in these organisms. If the organism is made polyploid, essentially doubling each of the chromosomes, every homolog will have a pairing partner and segregation will occur normally. Thus, fertility is restored.

TOPIC 2: ANEUPLOIDY

KEY POINTS

✓ *What is aneuploidy?*

✓ *What are the phenotypic consequences of aneuploidy?*

✓ *How does aneuploidy arise?*

All of the variations in chromosome number that were described in the previous topic were changes in the number of complete sets of chromosomes. All of those organisms are **euploid**, meaning they have complete sets of chromosomes. Note that *euploidy* does not specify the number of sets (i.e., the ploidy). Another class of chromosomal mutations includes those that change the numbers of single chromosomes. An organism that is one homolog shy of two complete sets is said to be **monosomic** (designated $2n - 1$ for two sets minus a single homolog). One that is missing all homologs of a particular chromosome is **nullosomic** ($2n - 2$). Those having a third homolog in addition to two sets are **trisomic** ($2n + 1$). All of these are examples of **aneuploidy** (*an-* means "not"), meaning that such organisms have incomplete sets of chromosomes. Aneuploidy is generally heritable if it does not interfere with viability or sex determination.

The first time aneuploidy was observed was in the experiments that followed the discovery of sex linkage in *Drosophila*. Morgan's student, Calvin Bridges, found more evidence for the chromosomal theory of inheritance in rare classes of progeny from Morgan's cross of a white-eyed female with a red-eyed male. Recall that the progeny of this cross were red-eyed females and white-eyed males. What was not described before was that about 1 of every 2,000 progeny was a white-eyed female or a red-eyed male. The red-eyed males (XO sex chromosomes) inherited their father's X chromosome carrying the red allele and no sex chromosome from their mothers. The white-eyed females (XXY sex chromosomes) inherited two X chromosomes from their mothers, each carrying a white allele, and a Y chromosome from their fathers.

In general, if an animal will tolerate aneuploidy of any sort, it will most likely be that of the sex chromosomes. This is probably due to the **dosage compensation** systems having evolved to manage sexually dimorphic chromosomes. For example, XXX human females inactivate both extra X chromosomes. No such mechanism is in place to manage incorrect dosage of autosomes. Consequently, autosomal aneuploidy often has serious phenotypic consequences. No autosomal monosomy is viable in humans. Monosomy of the X chromosome causes reduced viability and a characteristic phenotype called **Turner syndrome**. Only three autosomal trisomies result in live births, and two of these, trisomy 13 and trisomy 18, are associated with very short life expectancies. Trisomy 21 causes about 96% of the cases of a condition known as **Down syndrome**. As with the other trisomies, the Down syndrome phenotype is complex: short stature, mental retardation, characteristic facial features, abnormalities of internal organs, shortened life expectancy, and short, wide hands with special whorls, ridges, and folds in the skin. This complexity is a reflection of the number of genes affected by the mutation. The larger the chromosome, the more genes' dosages are increased by trisomy, and a complex set of phenotypes is produced; most human aneuploidies are lethal due to complex disruptions of the phenotype. Chromosome 21 is the smallest autosome, so presumably it contains fewer genes than most chromosomes. This may explain why Down syndrome patients often live into adulthood while none of the patients with other autosomal trisomies do; for a small chromosome compared to a larger chromosome, fewer genes would be present in three copies. A less severe disruption in gene dosage may cause a less severe phenotype, resulting in viability, rather than embryonic or infant death as does trisomy of any of the other autosomes.

What you now know about aneuploidy in humans is generally true of other mammals as well as birds. Yet, just as we saw for polyploidy, lower-order animals and plants are more tolerant of having too few or too many homologs of individual chromosomes. A striking example can be found in the plant genus *Datura* (jimsonweed). Diploid strains of this plant can be trisomic for any one of the organism's chromosomes and still be alive and fertile. Furthermore, each trisomy has a unique phenotype, reflecting the different genetic compositions of the individual chromosomes.

How do aneuploids arise naturally? Gametes are euploid because meiosis in most organisms causes homologs to synapse, and cross over, so that they can segregate away from each other. If a pair of homologs fails to synapse, or to cross over, segregation will not occur normally. Instead the homologs will behave independently at anaphase; sometimes they will move to opposite poles, giving euploid daughter cells, and sometimes they will move to the same poles, giving aneuploid daughter cells. **Disjunction** is the normal behavior of homologs at anaphase. Abnormal behavior, as described earlier, is termed **nondisjunction**. Nondisjunction can occur at meiosis I or meiosis II (**Figure 7.1**) and it is possible to distinguish between these two possibilities for a given trisomic individual if the parent in which nondisjunction occurs is heterozygous for a gene on the nondisjoined chromosome. Furthermore, nondisjunction can also occur during mitotic anaphase. Consequently, aneuploid offspring may arise despite normal gametes and due to mitotic nondisjunction in the zygote. To put these mutations in perspective, it has been estimated that aneuploidy occurs in roughly 5% of human conceptions.

Topic Test 2: Aneuploidy

True/False

1. Triploidy is the condition of having three homologs of only one chromosome.
2. The most commonly viable aneuploidy is that of the sex chromosomes.

Multiple Choice

3. Aneuploids have abnormal phenotypes because
   a. gene dosage is unbalanced.
   b. the individuals are sterile.

Normal chromosome disjunction

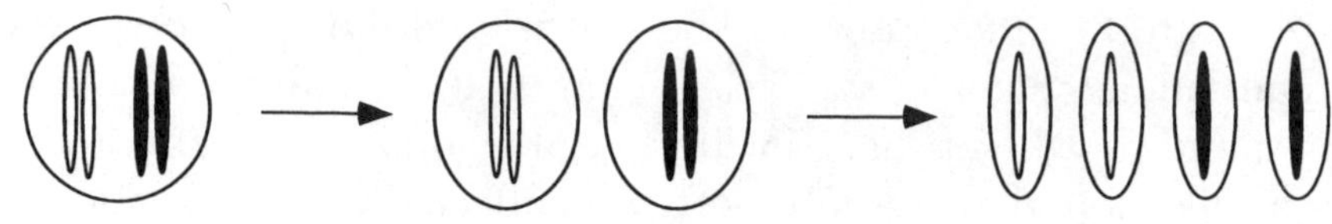

Nondisjunction in Meiosis I

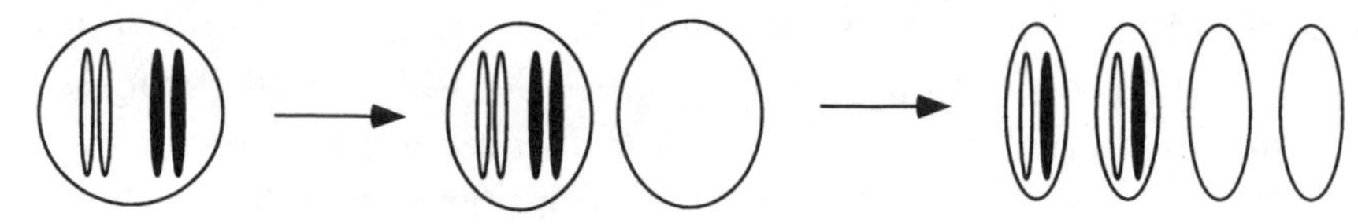

Nondisjunction in Meiosis II

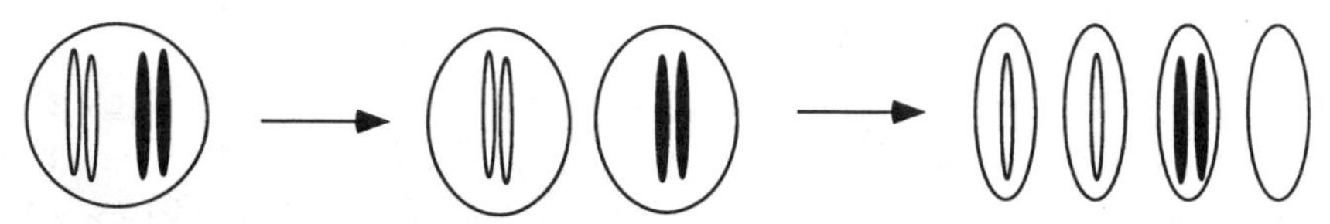

Figure 7.1 Nondisjunction of a single chromosome. Homologs are shaded. Meiosis I and meiosis II products are shown.

c. the sex chromosomes are affected.
d. of all of the above.
e. of none of the above.

4. Which of the following is *not* true of Down syndrome?
   a. If a Down syndrome individual is fertile, half of her offspring will have the syndrome.
   b. The complexity of the phenotype results from the large number of genes triplicated.
   c. The syndrome can be caused by nondisjunction in a parental meiosis.
   d. The syndrome can be caused by nondisjunction in zygotic mitosis.
   e. Most cases are caused by trisomy of chromosome 18.

Short Answer

5. A woman is heterozygous for the X-linked gene that causes recessive color blindness. Her husband is normal and they have a color-blind son. The son is karyotypically XXY. How can you explain his color blindness?

Topic Test 2: Answers

1. **False.** Triploidy is the condition of having three homologs of *every* chromosome. Trisomy is the condition of having three homologs of one chromosome.
2. **True.** Aneuploidy of the sex chromosomes causes the least aberrant phenotypes so is the most viable aneuploidy.
3. **a.** The phenotypes result from too much or too little of the gene products from the misrepresented chromosomes relative to the normal chromosomes.

4. **e.** The cause of Down syndrome is trisomy of chromosome 21.
5. If the son is color-blind despite having two X chromosomes, he must be homozygous for the color-blindness allele. That means both X chromosomes came from his mother because his father is normal. (It was probably nondisjunction in meiosis II that created the oocyte which contained two color-blindness X chromosomes.)

TOPIC 3: DELETIONS

KEY POINTS

✓ *What are the characteristics of deletions?*

✓ *What is deletion mapping?*

Deletion is one of the four types of structural aberrations of chromosomes. In Chapter 6, deletions were described as a type of point mutation—one that changes reading frame. Those deletions affected *single* genes. The deletions described in this section are the same in nature (i.e., removal of nucleotides), but very different in scope. By definition, chromosomal deletions remove the alleles of tens, or hundreds, or thousands of contiguous genes arrayed along a chromosome. The larger the deletion, the more genes are affected. Cytologically (microscopy of stained chromosomes) visible deletions affect huge numbers of genes. A deletion makes the cell containing it hemizygous for the affected genes; the only alleles these cells have for affected genes are the ones on the intact chromosome. For this reason, there can be no crossover between these genes in the meiosis of a deletion heterozygote. Consequently, the map distance will be changed in this region.

Large deletions are always heterozygous, never homozygous, because they invariably delete one essential gene or more (i.e., genes for which one allele is necessary for viability). Deletion heterozygotes may have a wild-type or a mutant phenotype. The terminal deletion of chromosome 5 in humans ("terminal" because the tip of the chromosome is deleted) produces a phenotype in the heterozygote called **cri du chat**. Named for the characteristic catlike cry of affected babies, the mutant phenotype results from an imbalance in gene dosage. Hemizygosity also means that any recessive mutant alleles on the normal chromosome that are missing from the deleted chromosome are not "covered" by dominant wild-type alleles; the recessive alleles will determine the phenotype of the organism because there will be no dominant alleles on the other chromosome to mask them. For this reason, deletions are said to "uncover" recessive alleles, especially recessive lethal alleles. This property can be used as a tool for mapping genes. A number of human genes, for example, were mapped to their respective chromosomes because some individuals with mutant phenotypes had a recessive allele of the relevant gene on one homolog and the other homolog for the same region was deleted. Such individuals are identified by their abnormal karyotypes, specifically a cytologically visible deletion, and it is concluded that the recessive allele resides in the same location on the intact homolog.

Deletion mapping is a technique that geneticists use to map genes to specific regions of chromosomes. In this technique, collections of strains of organisms, each having a different, mapped deletion (**Figure 7.2**), are mated to a strain carrying the mutation of interest. The genetic map in Figure 7.2 shows a portion of a hypothetical chromosome, with the relative locations of seven genes in a wild-type organism. The numbered strains contain the depicted deletion in a heterozygous state (because homozygosity may be lethal). A geneticist wishing to map a gene that

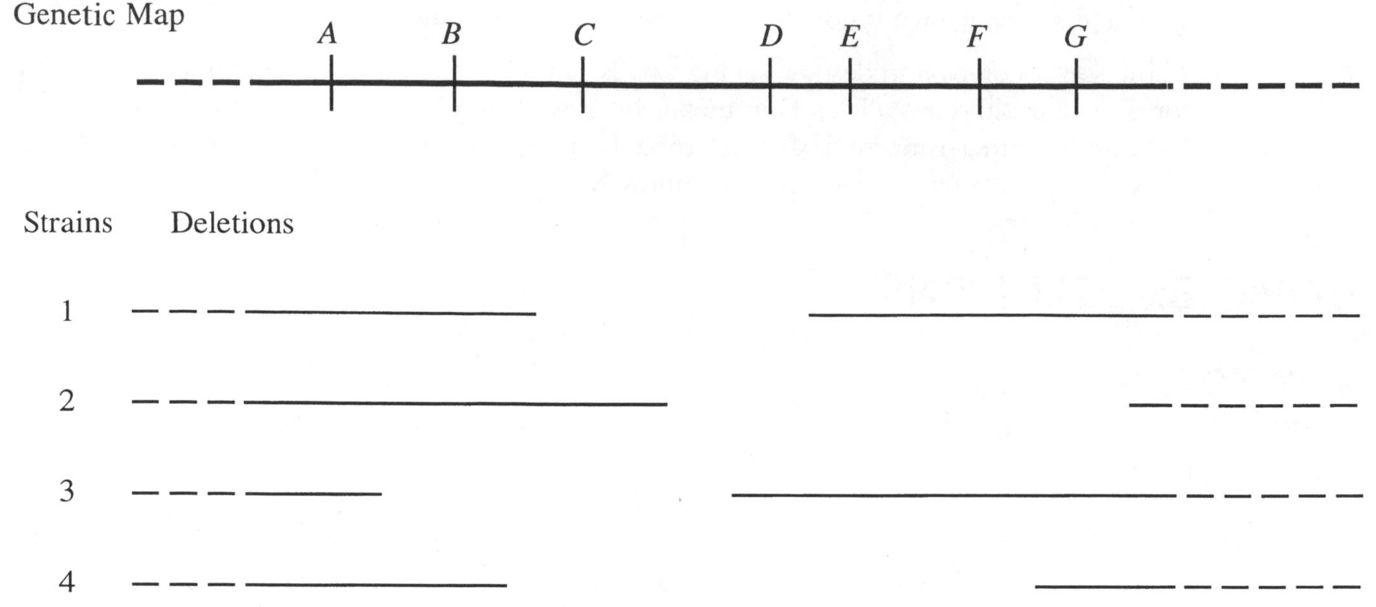

Figure 7.2 Deletion mapping. The dashed lines indicate continuation of the chromosomal map. Numbered strains contain the depicted deletion in a heterozygous state. The lines indicate DNA sequence remaining on the deleted homolog and the gaps represent the deleted sequence, compared to the wild-type map.

contains an interesting mutation crosses the mutant organisms to all of these (and more) deletion strains. The cross to strain 1 yields only wild-type progeny, as do the crosses to strains 2 and 4. However, the cross to strain 3 yields half wild-type and half mutant progeny. Only one gene that is deleted from strain 3 is present in all of the other strains: *B*. Therefore, the original mutant strain is homozygous for a mutation in gene *B* and gene *B* maps to this interval of the chromosome.

The precise mapping of relatively small deletions in model organisms is accomplished by crossing the deletion heterozygote to a multiply recessive organism, that is, performing a testcross of the deletion. Two kinds of progeny result. Half will have the normal chromosome from the deletion heterozygote and the recessive chromosome from the other parent. These will have a wild-type phenotype owing to the presence of the normal chromosome. The other half will show some of the recessive phenotypes. These second kind of progeny inherited the recessive chromosome from one parent and the deletion chromosome from the other parent. They will only show the recessive phenotypes of the genes that are missing from the deleted chromosome. This technique enables the researcher to determine the extent of the deletion.

Topic Test 3: Deletions

True/False

1. The size of the deletion is not related to its phenotype.
2. If a deletion is observable in the cells of an organism that has a non-wild-type phenotype, the gene(s) causing the phenotype may map to the region of the chromosome corresponding to the deletion.

Multiple Choice

3. Large deletions are likely to be
   a. homozygous wild type.
   b. heterozygous sterile.
   c. homozygous viable.
   d. homozygous lethal.
   e. none of the above.

4. Which one of the following characteristics is true of all deletions?
   a. Balanced gene dosage
   b. Wild-type phenotype in homozygotes
   c. Absence of crossover for genes within the deleted region
   d. Affects unlinked genes
   e. Can increase in size over sequential generations

Short Answer

5. Study the genetic map below. An organism that is heterozygous for a deletion mutation of this chromosome is crossed to a multiply recessive (but not deleted) individual, whose phenotype is abcdef. They have two kinds of progeny; half are wild type and half are phenotype bcde. Draw diagrams of the chromosomes present in the mutant progeny and label which parent contributed each chromosome.

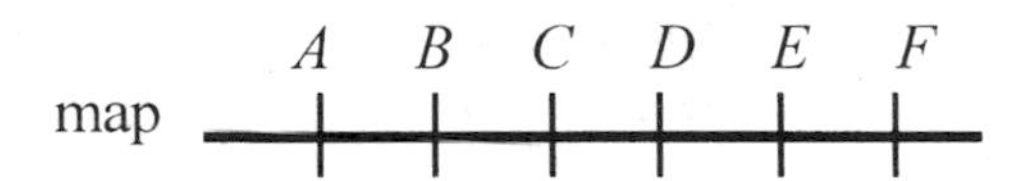

Topic Test 3: Answers

1. **False.** The larger the deletion, the more genes are likely affected, which generally makes the phenotype more complex.
2. **True.** This technique has identified the chromosomal locations of a number of genes responsible for human diseases.
3. **d.** The larger the deletion, the more likely it is to affect a gene that is essential for life.
4. **c.** Since the alleles of deleted genes are missing from that homolog, they cannot cross over with the intact homolog, and crossovers will not be observed in this region. For example, *ABCDE*/*ab* . . . *e* (the dots represent the deleted region) can cross over between *A* and *B* but not between *C* and *D*.
5. The dashed line in the drawing represents sequences missing from that chromosome.

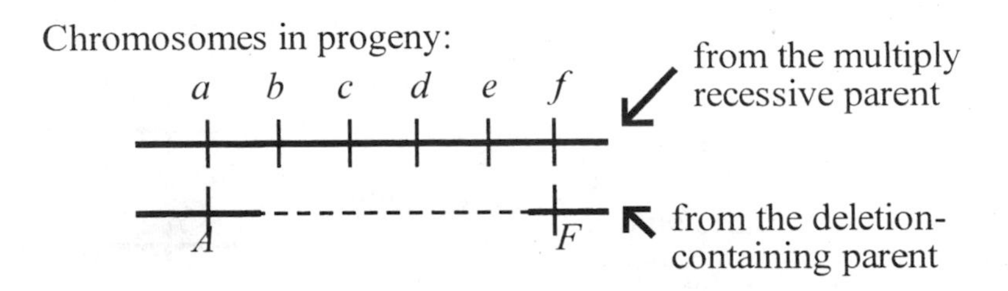

TOPIC 4: DUPLICATIONS

KEY POINTS

✓ *What are the general features of duplications?*

✓ *What role do duplications play in evolution?*

Duplication is the second type of chromosomal structural aberration to be discussed. It is easiest to picture this aberration as a duplication of a chunk of DNA, located at the site of the original gene(s), but the duplication *could* exist anywhere in the genome. Duplication of large parts or numbers of the chromosomes was important for the evolution of baker's yeast (*Saccharomyces cerevisiae*) and salmon (the Salmonidae family), to name just two examples. A typical, simple duplication can be depicted on a gene map as shown in **Figure 7.3**. In this map, genes *B* and *C* are duplicated. A heterozygote for the duplication would have three alleles each of genes *B* and *C*. A homozygote would have four alleles of each gene. In general, duplications are tolerated better than deletions, meaning they are more likely to cause a less serious phenotype or no phenotype at all. An example is the duplication that causes some cases of the human disorder Charcot-Marie-Tooth syndrome (CMT). The duplication is dominant, meaning heterozygotes for the duplication have the disorder. The CMT phenotype is progressive weakening of muscles and an associated loss of coordination. Phenotypes from duplications such as this one are caused by an imbalance in gene dosage. As pointed out earlier, the genes of many organisms exist in a balance and having too many or too few chromosomes or genes upsets this balance, resulting in complex phenotypes or death.

One special complication of duplications exists for future generations; once created, a duplication can continue to increase the number of alleles on the mutant chromosome by unequal crossover in meiosis. This is a rare event, but it is capable of creating triplications or higher-order reiterations of the original sequence. Furthermore, the unequal crossing over that increases the number of alleles occurs in all duplication genotypes except the homozygous wild type. The *Drosophila* Bar (small) eye phenotype is caused by a duplicated gene. Unequal crossing over in a homozygous *Bar* female can create a chromosome containing three alleles of this gene. The heterozygote possessing this triplicated chromosome and a normal chromosome has the same number of alleles as the homozygous *Bar* fly, and they have very similar phenotypes. Homozygotes for the triplication have even smaller eyes, indicating that eye shape is extremely sensitive to the dosage of this gene. The final point worth making about duplications pertains to evolution. Duplicating an essential gene allows one copy to retain the original function of the gene, while the duplicate is free to mutate randomly, perhaps acquiring a new function that is useful to the organism and causing the organism's divergence from related organisms. There is abundant evidence suggesting that this mechanism underlies the formation of many gene families in many species, including the globin gene families in vertebrates.

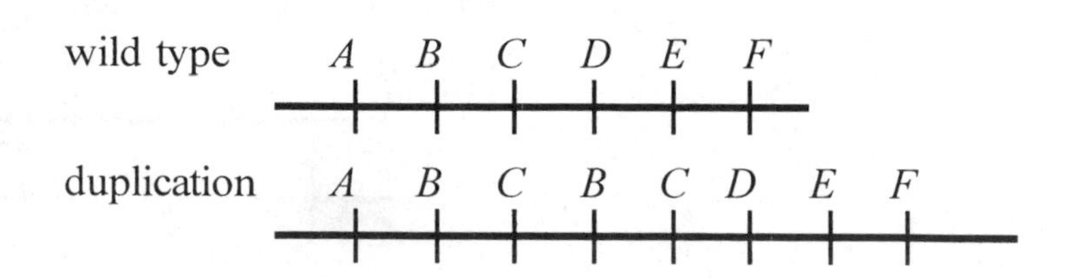

Figure 7.3 Genetic map showing a duplication.

Topic Test 4: Duplications

True/False

1. Any duplication might be dominant.
2. Duplications increase the numbers of chromosomes in cells.
3. Globin genes in humans were created through duplication of an ancient gene.

Multiple Choice

4. Which of the following statements is false?
   a. The size of duplications is variable.
   b. The location of duplications is variable.
   c. Duplications are tolerated better than deletions.
   d. Duplications can increase or decrease in number through unequal crossover.
   e. None of the above are false.

5. Duplications contribute to evolution by
   a. increasing the number of alleles to protect against divergence.
   b. increasing the number of alleles and allowing extra alleles to get new functions.
   c. decreasing surplus alleles to consolidate the genome.
   d. doing none of the above.
   e. doing all of the above.

Short Answer

6. Describe how gene duplication results in the Bar (small) eye phenotype in *Drosophila*. What do you think a (viable) homozygote for the quadruplication would look like?

Topic Test 4: Answers

1. **True.** Duplications increase gene dosage and might disrupt the balance of gene products. If they do upset the balance, they will be dominant.
2. **False.** Duplications increase the numbers of alleles of genes, ranging from one to many genes. A duplication could be as large as a chromosome but usually will not be.
3. **True.** Gene families such as the vertebrate globin genes are created by duplication and divergence through mutation.
4. **e.** All of the statements are true.
5. **b.** A duplication creates alleles that are not necessary for important processes. Mutation of these duplicate alleles can create new functions that may be beneficial.
6. Duplication of the *Bar* gene increases the amount of the *Bar* gene product, to which eye development is sensitive. Heterozygotes have smaller eyes than wild type, and homozygotes for the duplication have smaller eyes than heterozygotes. Homozygous triplication of the locus produces even smaller eyes than the homozygous duplication. By extension, the quadruplication would have the smallest eyes yet observed.

TOPIC 5: INVERSIONS

KEY POINTS

✓ *What are inversions?*

✓ *What are the recognizable features of inversions?*

✓ *How do inversions affect meiosis?*

Inversions are the third type of chromosomal structural aberration. These mutations change gene order (i.e., change the map) by flipping a portion of a chromosome relative to the rest. This is diagrammed in **Figure 7.4**. Note that no genes were lost or duplicated and that the gene pairs *B* and *E* and *D* and *A* are now closer to each other in the inversion than in the wild type. Generally, inversions are viable in both heterozygotes and homozygotes because they do not create an imbalance in genetic material. Furthermore, they often have wild-type phenotypes. The exception to this generalization is when one of the **breakpoints** of the inversion occurs within an essential gene. Such inversions are homozygous lethal.

The major problem for inversions is meiosis. Organisms bearing homozygous inversions undergo meiosis normally, with crossovers occurring normally, and the products reveal a rearranged genetic map relative to the wild-type map. Organisms that are heterozygous for an inversion have three distinctive features. First, the pairing of the normal and inverted homologs in heterozygotes shows a cytologically visible, diagnostic **inversion loop**. In order to synapse along their entire lengths, one homolog has to twist around the other, as shown in the right half of Figure 7.4. Second, there will be a decrease in the observed recombination frequency because crossovers within the inverted region create inviable products and therefore are undercounted. Third, a proportion of the gametes will be inviable, causing reduced fertility.

If the centromere is part of the inversion, the inversion is said to be **pericentric**. Crossovers outside the inverted region yield viable gametes, but there will be fewer recombinant progeny observed in the intervals immediately adjacent to the inversion. This occurs because parts of these intervals are included in the inversion, and they will not give viable crossover products. The amount that recombination is reduced is proportional to the amount of the interval included in the inversion; the more sequence between *A* and *B* (in Figure 7.4) that is not part of the inverted region, the more normal will be the recombination frequency of the *A*-to-inversion interval in the heterozygote. Crossover within a pericentric inversion causes one crossover product to be

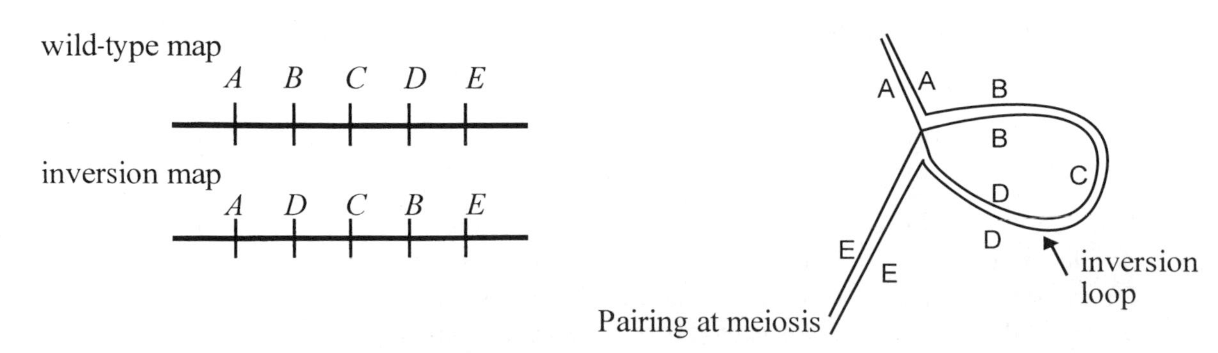

Figure 7.4 Gene map showing an inversion mutation and a diagram of an inversion loop.

duplicated for some genes and deleted for others, and the other product is the reverse. This creates a sizable proportion of progeny that are inviable due to an imbalance in gene dosage.

The other type of inversion is **paracentric**; this type does not include the centromere within the inverted sequence. Of the two products of a crossover *within* a paracentric inversion, one has two centromeres (**dicentric**) and the other has no centromere (**acentric**). The acentric fragment is lost because it cannot segregate. The dicentric fragment is pulled apart in anaphase I when the two centromeres are pulled toward opposite poles. This chromosome breaks at a random location between the two centromeres, creating two deletion chromosomes, one in each daughter cell. Both of these deletions are usually inviable.

Topic Test 5: Inversions

True/False

1. Gene dosage is upset by inversion mutations.
2. Inversions change the order of genes on the genetic map.

Multiple Choice

3. Which of the following is not a general feature of inversions?
   a. Homozygous lethality
   b. Fewer crossover products
   c. Reduced fertility
   d. Inversion loop
   e. None of the above
4. In what way are inversion heterozygotes compromised for meiosis?
   a. They do not cross over at all.
   b. They cross over outside the inversion but not within the inversion.
   c. They cross over normally but some of the products are inviable.
   d. They cross over too much everywhere.
   e. They cross over too much within the inversion.

Short Answer

5. Why do inversions normally have wild-type phenotypes? When are they not wild type?
6. Compare and contrast the products of crossover from a paracentric inversion heterozygote and a pericentric inversion heterozygote.

Topic Test 5: Answers

1. **False.** No genes are lost or duplicated so gene dosage is not changed.
2. **True.** This is a characteristic feature of inversions. If the chromosome had the arrangement *ABCDE*, an inversion might change that order to *ABEDC*.

3. **a.** Inversions are usually not homozygous lethal because most of the time they do not interrupt essential genes.

4. **c.** The machinery for crossing over is insensitive to the presence of an inversion. Some of the crossover products will have unbalanced gene dosage and that causes inviability.

5. Inversions normally have wild-type phenotypes because the breaks in the chromosome that cause the inversion rarely interrupt genes. Inversions are not wild type in the homozygous state when one of the ends of the inverted sequence was created by breaking the DNA within an essential gene. As long as the inversion is heterozygous, the normal chromosome can provide that gene product, giving the organism a wild-type phenotype.

6. A portion of progeny from both kinds of inversions are dead, as a result of unbalanced gene dosage. Crossovers in both inversions make deletion products. Duplication products come out of the pericentric inversion crossover but not the paracentric inversion heterozygotes. The paracentric inversion crossover creates aberrant (dicentric and acentric) chromosomes that go on to be deleted by the action of the segregation apparatus.

TOPIC 6: TRANSLOCATIONS

KEY POINTS

✓ *What are reciprocal and nonreciprocal translocations?*

✓ *What are the features of translocations?*

✓ *What effects do translocations have on organisms?*

Translocations are the fourth type of aberration in chromosome structure. A translocation is created when a piece of one chromosome is broken off, for example, by x-rays, and reattached to a nonhomologous chromosome. In the most common type of translocation, the second chromosome also loses a fragment and this fragment is reattached to the first chromosome. These translocations are said to be **reciprocal**. There is no loss of genetic material, so translocations usually have wild-type phenotypes, unless the reciprocal pair becomes separated (i.e., unbalanced) during meiosis or unless a breakpoint is within an essential gene (as with inversions). In these latter, uncommon circumstances, the phenotype is recessive lethality. The key feature of translocations is the radical change they produce in the genetic map. Some genes that are unlinked in the wild type are linked by the translocation; some that are linked in the wild type become unlinked by the translocation. In the wild-type chromosomes represented in **Figure 7.5**, *E* and *W* show independent assortment, but in the reciprocal translocation *E* and *W* are linked. This rearrangement of genes will not affect mitosis, but heterozygotes will be compromised for meiosis. The homologous sequences will synapse completely if the normal and translocated chromosomes assume a cruciform (crosslike) configuration like the one shown in **Figure 7.6** for the previously depicted chromosomes. Crossover frequencies for these chromosomes are normal, given the new linkage relationships. If the offspring inherit both normal (N) chromosomes or both translocated (T) chromosomes, they will be phenotypically normal. This disjunction is termed **alternate segregation**. If the disjunction is **adjacent segregation**, the gametes contain one T and one N chromosome, and the resulting offspring will have duplications of some genes and deletions of others. These are usually inviable or have a severe phenotype.

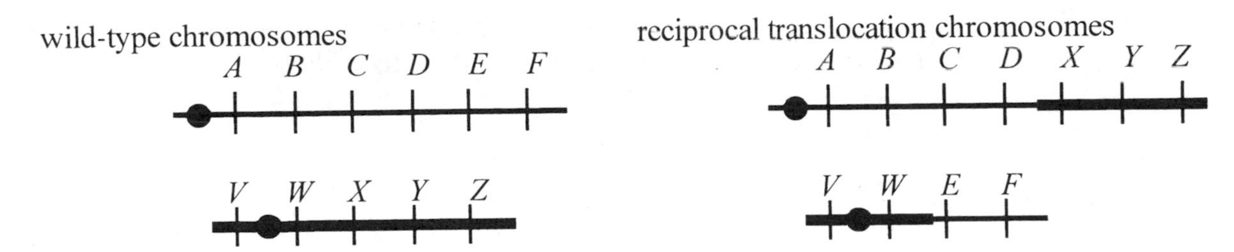

Figure 7.5 Reciprocal translocation. Dots are centromeres.

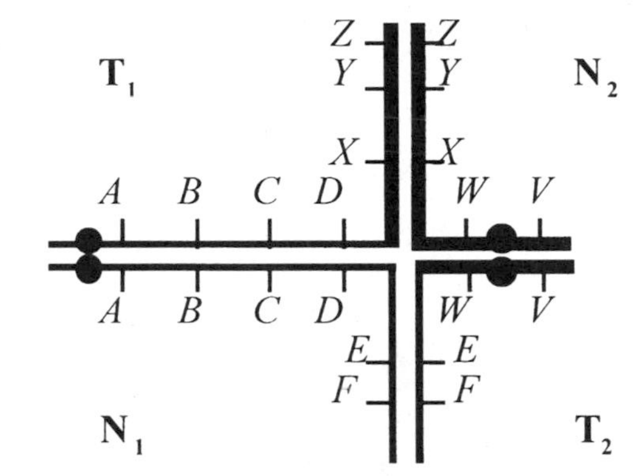

Figure 7.6 Meiotic pairing configuration of the translocation heterozygote represented in Figure 7.5.

The inherited form of Down syndrome in humans is an example of this semiviable, phenotypic severity. About 4% of Down syndrome patients have (in addition to one normal homolog of chromosome 14 and two of chromosome 21) a nonreciprocal translocation of one chromosome 21 homolog onto the end of a chromosome 14 homolog, resulting in a very large chromosome that contains the chromosome 14 centromere and most of the material of both chromosomes. Heterozygous carriers of the translocation have one normal homolog each of chromosomes 14 and 21 plus the large translocation chromosome, giving them an abnormal number of chromosomes, 45. These heterozygotes make six kinds of gametes, not at equal frequency. The two gametes that result in phenotypically normal progeny occur when the translocation disjoins from the two normal chromosomes (an alternate segregation). These contain either both normal chromosomes 14 and 21 or the translocation chromosome, and the resulting offspring have the correct number of copies of each chromosome, although one is heterozygous for the translocation. Another (adjacent) segregation occurs when chromosome 21 segregates with the translocation rather than with chromosome 14 and creates gametes that have either chromosome 14 (but not chromosome 21) or both chromosome 21 and the translocation. The first gamete can only produce inviable zygotes, owing to chromosome 21 monosomy. The second gamete is fertilized to give an offspring that has two normal copies of chromosome 21, one normal chromosome 14, and the translocation, which adds a second dose of chromosome 14 and a third dose of chromosome 21. It is the third dose of chromosome 21 in this offspring that causes the Down syndrome phenotype.

The third (adjacent) segregation is very rare because the translocation chromosome has the chromosome 14 centromere, and the remaining segregation pattern requires these two homologous centromeres to segregate together, instead of separating normally. The resulting gametes give inviable zygotes. Among the viable offspring, the translocation heterozygote has three classes of children: those who are phenotypically and genotypically normal, those who are phenotypically normal and are heterozygous for the translocation, and those who have Down syndrome. The translocation heterozygote has a relatively high probability (approaching 1/3) of each child having Down syndrome. Contrast this with the situation for most Down syndrome cases, caused by trisomy 21, where the risk per child is roughly 3/2,000. In addition, the heterozygous progeny of the heterozygous carrier are also at relative high risk of having affected offspring; that is, the relatively high risk of affected progeny is a heritable trait.

In summary, translocations alter the genetic map, homozygotes occasionally have a non-wild-type phenotype, and heterozygotes have both decreased fertility due to segregation errors and some non-wild-type offspring.

Topic Test 6: Translocations

True/False

1. If a break that occurred to create the translocation disrupted an essential gene, the translocation will be recessively lethal.
2. Translocations are reciprocal if two nonhomologous chromosomes exchanged segments of DNA with each other.
3. Crossover within a translocation creates inviable gametes.

Multiple Choice

4. Which one of the following is not a general feature of translocations?
   a. Reciprocal translocation heterozygotes exhibit cruciform-shaped pairing.
   b. If both of the reciprocal translocation chromosomes are present, the phenotype is usually wild type.
   c. If the translocation is reciprocal, the offspring that inherit only one of the translocated chromosomes are inviable.
   d. Translocations that are not reciprocal are not wild type in phenotype.
   e. All of the above are general features.
5. What sorts of progeny can carriers of the Down syndrome translocation have?
   a. Genotypically normal progeny
   b. Down syndrome progeny
   c. Genotypically normal progeny and Down syndrome progeny
   d. Genotypically normal progeny and carriers of the translocation
   e. Genotypically normal progeny, carriers of the translocation, and Down syndrome progeny

Short Answer

6. Draw the pairing configuration for the Down syndrome translocation and indicate which segregation produces the "affected" gamete.

Topic Test 6: Answers

1. **True.** Heterozygotes will be viable because the normal chromosome can supply the essential gene product. Only homozygotes for this translocation would die.

2. **True.** The alternative, one chromosome donating DNA to another and gaining nothing in return, is not reciprocal.
3. **False.** Crossover is normal and creates viable products; it is the way the chromosomes are segregated that can cause lethality.
4. **d.** Most translocations give wild-type phenotypes regardless of whether they are reciprocal.
5. **e.** Alternate segregation produces genotypically normal progeny and carriers of the translocation. Adjacent segregation produces the progeny who have Down syndrome.
6. This is an adjacent segregation.

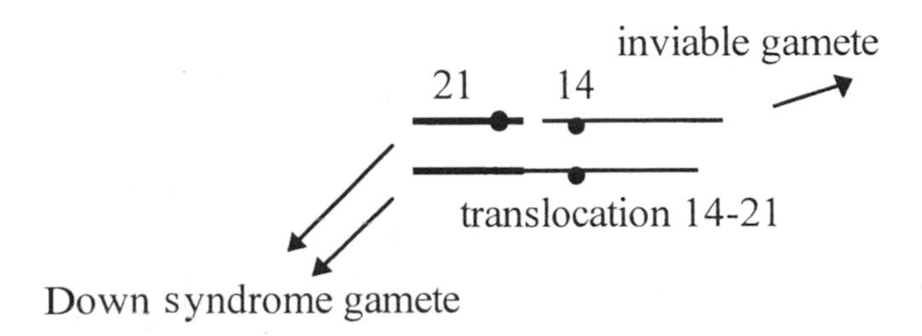

IN THE CLINIC

Chromosomal aberrations are a significant cause of human infertility. It has been estimated that about 8% of human conceptions involve chromosomally abnormal zygotes. Most of these zygotes are inviable as a result of the mutations and are spontaneously aborted. These abortions occur, by definition, within the first 22 weeks of the pregnancy or when the embryo's weight is less than 500 grams. Practically speaking, most of the spontaneous abortions that are caused by chromosomal abnormality occur during the first 2 months of gestation. Later abortions are more likely to result from less severe disturbances in gene dosage. The rates of chromosomal abnormalities and spontaneous abortions are estimates because many of these abortions occur before the woman knows she is pregnant (within the first 2 weeks of conception) or without examination of the aborted fetus. Approximately 1 of every 170 live births is of a child who has an obvious chromosomal abnormality. Polyploidy, primarily triploidy, accounts for 1% to 2% of conceptions and is always lethal. The incidence of particular trisomies correlates with maternal age or paternal age. Notably, the frequency of trisomy 21 (Down syndrome) correlates with maternal age. A 20-year-old woman has nearly a 1/2,000 risk of a trisomy 21 fetus, while a 40-year-old woman has roughly a 1/100 risk. There is no correlation with paternal age for Down syndrome; however, other chromosomes are nondisjoined at increased frequency in older males. In short, each chromosome has age-dependent and sex-dependent nondisjunction rates. Various models have been proposed to explain the age dependence of chromosome 21 nondisjunction in females specifically, but the cause is unknown.

Chapter Test

True/False

1. Aneuploid organisms are generally fertile.
2. Duplications can increase their number of reiterations by crossing over.
3. Inversion heterozygotes have reduced crossover during meiosis.
4. Translocations do not alter the genetic map.
5. Large deletions are homozygous lethal.
6. Translocations are most likely to cause phenotypes when unbalanced.
7. In general, duplications are tolerated better than deletions are.

Multiple Choice

8. Which one of the following statements is not true of deletions?
   a. Deletion removes one allele of each affected gene.
   b. Deletions may be cytologically visible.
   c. Deletions prevent crossover between affected genes.
   d. Deletions compromise meiosis.
   e. All of the above are true.

9. What is trisomy?
   a. The presence of three alleles of a gene
   b. The presence of three homologs of a chromosome
   c. The presence of three sets of chromosomes
   d. The presence of three ears
   e. None of the above

10. Phenotypes of duplications are
    a. wild type or dominant.
    b. wild type or recessive.
    c. dominant.
    d. wild type.
    e. recessive.

11. Consider the reciprocal translocation heterozygote whose meiotic pairing configuration is shown in Figure 7.6. There are several gametes that can result from this pairing. Which of the following are not likely?
    a. N_1N_2
    b. T_1T_2
    c. N_1T_1
    d. N_1T_2
    e. More than one of the above

12. The general phenotype of polyploids, compared to their diploid relatives, is
    a. wild type.
    b. bigger.

c. more fertile.
d. deformed.

13. What progeny phenotypes are *not possible* from the testcross of a paracentric inversion heterozygote whose homologs are *.ABCDE* and *.aedcb* (the dots represent the centromere).
    a. Wild type
    b. a
    c. abcde
    d. bcde
    e. cde

Short Answer

14. What events result in polyploidy?
15. A human who has three sex chromosomes, XYY, is a fertile male. What are the possible sex chromosome constitutions of the gametes he makes?
16. The normal gene order for a particular chromosome is *MNOP*. Predict the progeny's phenotypes from the testcross of an individual who is (a) heterozygous for a deletion of *N* and *O*; (b) heterozygous for a duplication of *N* and *O* (gene order *MNONOP*, and assume the duplication's phenotype is dominant); and (c) heterozygous for an inversion of *N* and *O* (gene order *MONP*).

Essay

17. Of the four structural aberrations, which have trouble in meiosis? For each of the four types, describe when they have non-wild-type phenotypes.

Chapter Test Answers

1. **T** 2. **T** 3. **F** 4. **F** 5. **T** 6. **T** 7. **T** 8. **d** 9. **b** 10. **a** 11. **c** 12. **b**

13. **e**

14. Nondisjunction of all of the chromosomes during mitosis (followed by vegetative propagation) or during meiosis (followed by fertilization of the polyploid gamete).

15. X, YY, XY, Y

16. (a) The deletion heterozygote is *MP/MNOP*, which is crossed to *mnop* to give 1/2 wild type and 1/2 *no* progeny. (b) The duplication heterozygote is *MNONOP/MNOP*, which is crossed to *mnop* to give 1/2 dominant and 1/2 wild-type progeny. (c) The inversion heterozygote is *MONP/MNOP*, which is crossed to *mnop* to give all wild-type progeny.

17. Cells with translocations and inversions have compromised meioses. Translocations can disjoin in several ways, some of which will cause severe or lethal phenotypes in the offspring. Inversions create deletion and duplication products that can be lethal in the offspring. Deletions cause phenotypes in homozygotes and sometimes in heterozygotes. Duplications can cause phenotypes in heterozygotes or homozygotes. Inversions rarely have phenotypes, except that heterozygotes have reduced fertility. Reciprocal translocations

cause non-wild-type phenotypes when they become unbalanced through adjacent segregation in a heterozygote. Translocations of either type (reciprocal or not) can cause reduced fertility in heterozygotes. Inversions or translocations can cause phenotypes if the breakpoints fall in essential genes.

Check Your Performance:

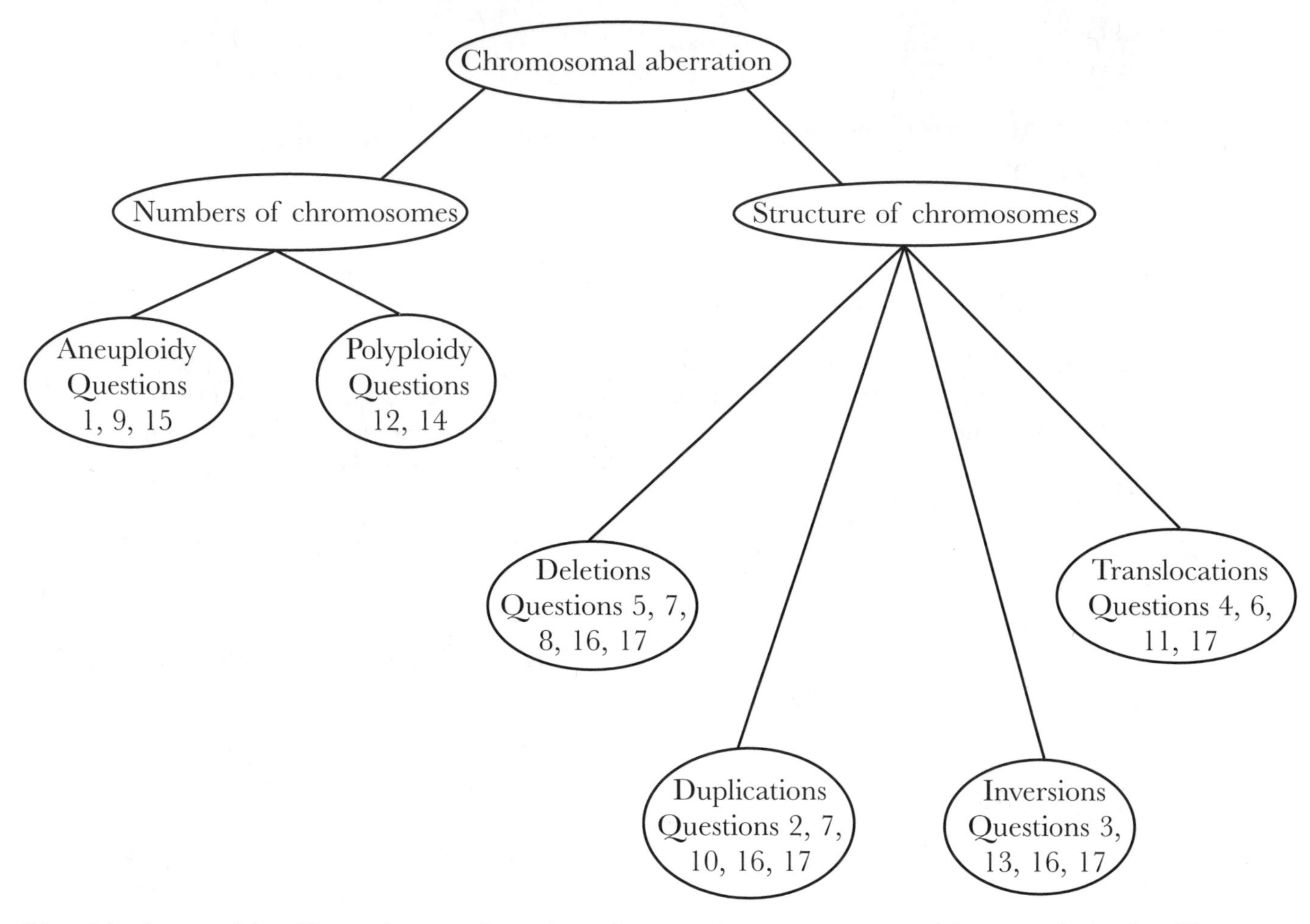

Use this chart to identify weak areas, based on the questions you answered incorrectly in the Chapter Test.

Second Midterm Exam

True/False

1. Most species are capable of being polyploid.
2. One crossover produces two recombinant progeny.
3. Mutagens are substances that increase the rate of mutation.
4. Semiconservative replication results in one helix being composed of old strands and one being composed of new strands.
5. Selfed dihybrids can not show linkage.
6. Error-prone repair decreases the rate of DNA polymerase errors.
7. DNA exists within the nucleus in a complex with proteins.
8. Polyploidy can arise spontaneously or can be induced.
9. Deletions make diploid cells hemizygous for the affected genes.
10. A point mutation is the change of one nucleotide to another.
11. Map distances are calculated from the percent of parental-type progeny.
12. Three rRNA genes are transcribed into one precursor molecule.

Multiple Choice

13. In what way do translocations change the genetic map?
    a. Some loci move closer together and some farther apart.
    b. Crossover causes reiteration of some sequences.
    c. Some unlinked genes become linked.
    d. There is an absence of crossover within a particular region.
    e. Crossover within a particular region produces inviable gametes.
14. Which of the following is not a feature of most DNA?
    a. Antiparallel
    b. Complementary
    c. Double stranded
    d. Hydrogen bonding
    e. Positive charge
15. Spontaneous mutations are
    a. less deleterious than induced mutations.
    b. adaptive.
    c. somatic.
    d. induced by mutagens.
    e. the result of normal cellular activities.
16. Imagine a cross between a *cis* dihybrid and a double homozygous recessive in which the segregating genes are extremely closely linked. What phenotypes do you expect in 30 progeny?

a. All dominant
b. All recessive
c. 1/2 dominant and 1/2 recessive
d. 1/4 each of four classes
e. Four classes in a 9:3:3:1 ratio

17. Which of the following is not a characteristic of DNA?
    a. Major groove
    b. carboxyl terminus
    c. 2-nm-diameter helix
    d. 10-nm fiber
    e. 30-nm fiber

18. Which one of the following statements is true?
    a. All duplications cause phenotypes.
    b. Any duplication might cause a phenotype.
    c. Duplications cause dominant phenotypes.
    d. Duplications cause recessive phenotypes.
    e. Crossover is suppressed in duplications.

19. DNA polymerase is responsible for removing
    a. mismatched nucleotides.
    b. pyrimidine dimers.
    c. x-rays.
    d. chemically modified bases.
    e. error-prone synthesis.

20. How are dicentric chromosomes created?
    a. By crossover within the inverted region of a pericentric inversion heterozygote
    b. By crossover outside the inverted region of a pericentric inversion heterozygote
    c. By crossover within the inverted region of a paracentric inversion heterozygote
    d. By crossover outside the inverted region of a paracentric inversion heterozygote
    e. By unequal crossover between duplications

Short Answer

21. What phenotypes do you expect in the progeny of a testcross of a deletion heterozygote? The heterozygote's chromosomes are *ZYXWV* and *zy . . . v*. Include progeny that could result from crossover.

22. A fictional cross involves squirrels who are heterozygous for three loci: fur color (*R* is brown and *r* is red), whisker color (*S* is silver tips and *s* is solid), and fur smoothness (*F* is puffy and *f* is flat). Testcrosses of these heterozygotes yield 400 progeny, by the following phenotypes:

F R S	71
f r s	72
F R s	73
f r S	70
F r s	27

f R S	29
F r S	28
f R s	30

Draw a map for the linked genes and include distances.

23. Genes *A* and *B* are 5 map units apart and genes *B* and *C* are 10 map units apart. (The order is *ABC*). What is the expected frequency of double crossover?

Essay

24. Why are frameshift mutations more likely than base-substitution mutations to result in proteins lacking function?

25. Describe initiation and termination of translation in eukaryotes.

26. In what way are polyploids agriculturally important?

Answers

1. **F** 2. **T** 3. **T** 4. **F** 5. **F** 6. **F** 7. **T** 8. **T** 9. **T** 10. **F** 11. **F** 12. **T**

13. **c** 14. **e** 15. **e** 16. **c** 17. **b** 18. **b** 19. **a** 20. **c**

21. There will be crossover between *Z* and *Y* and between *Y* and the deletion breakpoint and between the deletion breakpoint and *V*. Progeny can be ZYXWV and zyxwv (the parental types), Zyxwv, zYXWV, ZYxwv, zyXWV, ZyxwV, zYXWv, ZYXWv, zyxwV, zyXWv, and ZYxwV.

22. There appear to be four parental classes, instead of two, indicating independent assortment of the whisker color gene.

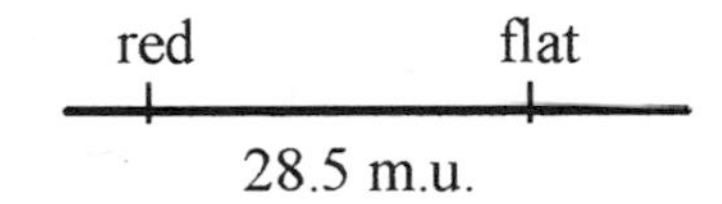

23. The expected frequency of double crossover is 0.005, which is obtained by multiplying the frequency of crossover in the two intervals: 0.05×0.10.

24. Frameshift mutations alter the entire protein sequence that follows the site of the mutation. Base-substitution mutations generally only change a single amino acid in the protein, provided the mutation is not silent. Of the two remaining consequences of base substitution, single missense mutations often will not completely disrupt protein function. Only nonsense mutations are very likely to cause proteins that lack function. Consequently, a single altered nucleotide (i.e., a single altered codon) is less likely than a frameshift mutation (i.e., many altered amino acids) to disrupt protein function.

25. A complex containing the small ribosome subunit binds the mRNA and an initiator tRNA and searches for the AUG initiation codon in the 5′ end of the transcript. When this is located, the large subunit binds and initiation is complete. Termination occurs when a stop codon, bound by release factors, reaches the A site of the ribosome.

26. As ploidy increases, size generally does so, as well. Secondly, when fertility interferes with marketability, as in the case of large seeds, triploidy will prevent or greatly reduce the number of seeds produced. Thirdly, in the breeding of new varieties, hybrids between closely related species are often sterile. Increasing the ploidy of such organisms can make them fertile, provided they tolerate polyploidy.

The Gene as a Functional Unit

Today we know that genes encode proteins and protein activity determines phenotype. However, in the first half of this century, the relationship between genes and proteins was not clear. Mendel showed that heritable factors determine phenotypes, but he did not know the basis for this relationship. A series of elegant experiments clarified the matter and led to the *central dogma*. [Remember that the central dogma states that genes (DNA) are copied into mRNA, which is used to make protein.] Some of the experiments critical to the understanding that genes encode proteins are described in this chapter.

ESSENTIAL BACKGROUND

- **Dominance and recessiveness (Chapter 1)**
- **Incomplete dominance, gene interactions, and their ratios (Chapter 3)**
- **Recombination frequency (Chapter 4)**
- **Gene structure and protein structure (Chapter 5)**

TOPIC 1: ONE GENE–ONE ENZYME

Key Points

✓ *What is the proof that genes affect enzymes?*

✓ *What is the one gene–one enzyme hypothesis?*

✓ *How is sequential gene function interpreted in biochemical pathways?*

Two discoveries led to the understanding that genes and enzymes are related in a specific manner. The first discovery was made in 1902 by the English physician Archibald Garrod, who described the genetic inheritance of a metabolic defect. The disease he observed was alkaptonuria, which causes the urine of affected individuals to turn black when exposed to air. Garrod recognized recessive inheritance of the trait in these families; the disease clustered in a few families and was especially likely among the children of first cousins. He also determined the identity of the unusual chemical in the alkaptonurics' urine: homogentisic acid. Garrod inferred that normal individuals metabolize this chemical into another substance and that alkaptonurics cannot because they lack a particular enzyme that is required for metabolism (degradation) of homogentisic acid. Garrod's study of alkaptonuria and three other diseases contributed to his theory that genetic defects can disrupt metabolic pathways. He called these diseases **inborn errors of metabolism**.

The relationship between genes and enzymes was confirmed in 1942 by George Beadle and Edward Tatum, who received a Nobel Prize for this work. Their experiments used a fungus, *Neurospora crassa*, which has a haploid genome and therefore could easily be mutagenized to auxotrophy. (**Auxotrophy** means these mutant strains could no longer grow on minimal medium; they had to be supplemented with one or more nutrients.) To be certain they were studying simple genetic mutations, Beadle and Tatum tested each of the individual strains to prove the mutations were inherited as single gene defects. Among their mutant strains were a set that could not grow on minimal medium without a supplement of the amino acid arginine (hence, arg^- mutants). Three strains whose mutations mapped to different chromosomes and therefore represented separate genes were selected for further study.

Interestingly, the strains differed in the nature of their defects. Other researchers had previously determined that cellular enzymes can convert related compounds into each other. So Beadle and Tatum tested their strains for their abilities to grow on several compounds related to arginine. One strain (let us call it *arg-1*) could grow on minimal media that had been supplemented with arginine, ornithine, or citrulline. A second strain, *arg-2*, could grow on minimal media that had been supplemented with arginine or citrulline but not with ornithine. A third strain, *arg-3*, could only grow in the presence of arginine. Citrulline and ornithine were not capable of rescuing the growth defect of the *arg-3* strain.

These phenotypes suggested to Beadle and Tatum both a metabolic pathway for these compounds and a role for each gene in that pathway. Their explanation was the **one gene–one enzyme hypothesis**. It states that each gene controls the activity of a specific enzyme. Each of Beadle and Tatum's *arg* mutations identifies a gene that encodes one of the enzymes required for arginine biosynthesis. Each enzyme converts a metabolic precursor into another component in the biosynthetic pathway.

The *arg-1* mutant, carrying a specific mutation in a single gene, is missing an enzyme that is necessary for the synthesis of arginine. The strain containing this mutation cannot grow in the absence of arginine. The same is true of *arg-2* and *arg-3* mutants. We know that these genes encode different enzymes because these mutants are blocked at different points with respect to metabolic precursors. The *arg-1* mutants can grow on ornithine and citrulline and arginine. This suggests that *arg-1* acts before, or "upstream" of, these components in the biosynthetic pathway; all three chemicals bypass the "broken" step in the pathway. Having these chemicals available in the medium means the organism does not have to make them itself, so if it cannot make them because of a mutation, the chemicals' availability in the medium allows growth.

In contrast to the *arg-1* mutant, an *arg-2* mutant cannot grow on ornithine, suggesting that the enzyme encoded by *arg-2* acts after ornithine in the pathway. The *arg-2* gene product is required to convert ornithine, ultimately, to arginine, so even if ornithine is supplied, without wild-type *arg-2* gene product, the organism cannot grow. By contrast, citrulline allows *arg-2* mutants to grow because it, and its conversion to arginine, lie downstream of the mutation (or the "block"). This ability of *arg-2* mutants to use a substance that *arg-1* mutants cannot means that *arg-2* lies downstream of *arg-1* in the pathway.

The *arg-3* mutants are incapable of growing in the presence of ornithine or citrulline, suggesting the biosynthetic pathway is blocked at a point after these intermediates, but before arginine.

Hence, the *arg-3* gene lies downstream of both *arg-1* and *arg-2*, because this mutant is able to use the fewest substances for growth. It is possible to depict this explanation as a pathway, where arrows lead from precursors to products and genes label each arrow to indicate the gene product catalyzing each step. Note that for each biosynthetic pathway you will see, the final substance generated by the pathway is necessary for the organism to have the wild-type phenotype.

$$\text{precursor} \xrightarrow{\textit{arg-1}} \text{ornithine} \xrightarrow{\textit{arg-2}} \text{citrulline} \xrightarrow{\textit{arg-3}} \text{arginine}$$

The modern restatement of this observation is one gene–one polypeptide because we now know that enzymes can be complexes of two or more polypeptides and that some polypeptides have more than one function or activity.

Topic Test 1: One Gene–One Enzyme

True/False

1. Biosynthetic pathways show the enzymes and intermediates that produce substances that are necessary for the wild-type phenotype.
2. The one gene–one enzyme hypothesis states that each gene controls the activity of a specific enzyme.

Multiple Choice

3. This pathway shows the degradation of the amino acid tyrosine, where each arrow represents the activity of a different enzyme. Note that this differs from a biosynthetic pathway in that the end product is excreted.

$$\text{tyrosine} \longrightarrow p\text{-hydroxyphenylpyruvate} \longrightarrow \text{homogentisic acid} \xrightarrow{\textit{genX}}$$
$$\text{maleylacetoacetate} \longrightarrow \text{fumarylacetoacetate} \longrightarrow \text{acetoacetate}$$

What is the phenotype of an organism who is mutant for *genX*?
a. It excretes acetoacetate.
b. It excretes fumarylacetoacetate.
c. It excretes maleylacetoacetate.
d. It excretes homogentisic acid.
e. It excretes *p*-hydroxyphenylpyruvate.

4. In the pathway shown below, the indicated genes specify enzymes that synthesize pigmented substances. A mysterious organism uses this pathway to make dark fur. A rare individual, however, is blue. How can you explain this oddity?

$$\text{colorless} \xrightarrow{\textit{enzA}} \text{white} \xrightarrow{\textit{enzB}} \text{blue} \xrightarrow{\textit{enzC}} \text{black}$$

a. The individual cannot make the white pigment.
b. The individual cannot make the blue pigment.
c. The individual cannot make the black pigment.

d. The individual converts black pigment to blue pigment.
e. The individual converts blue pigment to white pigment.

Short Answer

5. Unlike most organisms, some bacteria can synthesize vitamin B_{12}, which is necessary for growth in all organisms. Bacteria that are mutant for one of a few genes cannot synthesize this vitamin from simple molecules but can convert one or more related compounds into the vitamin. Mutation of "*gene1*" results in bacteria that can grow on minimal media supplemented with vitamin B_{12}, but no other compound will support growth. Mutation of "*gene2*" results in bacteria that can grow when supplemented with cobinamide, cobyrinate, porphobilinogen, or vitamin B_{12}. Mutation of "*gene3*" results in bacteria that can grow when supplemented with cobinamide or vitamin B_{12}. Mutation of "*gene4*" results in bacteria that can grow when supplemented with cobinamide, cobyrinate, or vitamin B_{12}. What is the pathway for vitamin B_{12} synthesis? Show both intermediates and genes.

Topic Test 1: Answers

1. **True.** Enzymatic pathways exist so that organisms can obtain the molecules they need for growth when those molecules are not available in their environments. The outcomes of biosynthetic pathways are essential for the organism to have a wild-type phenotype (or to survive in particular circumstances). Other pathways, such as the one containing homogentisic acid, are for degradation and the products are not necessary for growth, but are wild type.
2. **True.** Mutations in single genes manifest in the lack of activity in specific enzymes. This was the simplest conclusion from the studies of Garrod and of Beadle and Tatum.
3. **d.** Mutants in *genX* will lack the enzyme that converts homogentisic acid to maleylaceto-acetate. Consequently, they will excrete homogentisic acid. This is the mutant gene studied by Garrod.
4. **c.** Normal individuals convert the blue pigment to the black pigment. The organism must be mutant for *enzC* so that blue pigment accumulates.
5. Vitamin B_{12}, which is essential for growth, is the end product of this pathway. The *gene1* mutants grow only when supplemented with the end product so this gene product must act near the end of the pathway. *Gene2*, *gene3*, and *gene4* lie upstream of *gene1*. The *gene3* mutants grow when supplemented with two compounds; *gene4* mutants, with three compounds; and *gene2* mutants, with four. This is the reverse order of their action. The final pathway shows *gene2* acting first, to produce the compound that no other mutation can use, porphobilinogen. *Gene4* acts downstream of *gene2*, producing the second "least used" compound, cobyrinate. *Gene3* acts next, producing the compound (cobinamide) that is used almost as commonly as the end product, vitamin B_{12}.

$$\text{minimal medium} \xrightarrow{gene2} \text{porphobilinogen} \xrightarrow{gene4} \text{cobyrinate} \xrightarrow{gene3} \text{cobinamide} \xrightarrow{gene1} \text{vitamin } B_{12}$$

TOPIC 2: RELATIONSHIP BETWEEN GENE SEQUENCE AND PROTEIN SEQUENCE

KEY POINTS

- ✓ *What is the proof that genes and proteins are collinear?*
- ✓ *How do genetic observations relate to gene function and protein function?*

The first demonstration that mutational changes in genes could alter protein sequence came in 1957 from Vernon Ingram. Ingram studied the blood protein hemoglobin, which transports oxygen. Sickle-cell anemia is a serious, heritable disorder characterized by misshapen red blood cells in homozygotes. Ingram found that a fragment of hemoglobin differs in people affected with sickle-cell anemia versus normal individuals. He sequenced the aberrant polypeptide and found it to have a valine at position 6, a change from the glutamic acid that is found at this position in the normal polypeptide. People who are homozygous for the sickle-cell mutation (a simple base-substitution mutation that alters a single amino acid) make an aberrant polypeptide that causes a complex phenotype consisting of many internal organ malfunctions that eventually result in premature death. This is a severe consequence for a simple mutation. Ingram did not know the DNA nature of the mutation (the gene's sequence was not known); otherwise, he might have gone on to prove that gene sequence directly specifies protein sequence.

Charles Yanofsky added the crowning touches to this understanding of the gene-protein relationship by proving that genes and proteins are collinear; that is, their linear sequences correlate. He studied mutations in the *trpA* gene in *Escherichia coli* and was able to order the mutations on a map by linkage analysis. When he determined the sequences of the proteins encoded by the mutant alleles, he found a direct correlation between the order of mutations on his map and the positions of the corresponding altered amino acids. Mutations that mapped to the 5′ end of the gene affected amino acids in the N-terminal end of the protein. Mutations that mapped to the 3′ end of the gene affected amino acids in the C-terminal end of the protein. When two mutations were close together, the altered amino acids they produced were also close together. This was the first clear evidence that the sequence of a gene determines the sequence of a protein. It is particularly impressive evidence in that it was created without knowledge of the DNA sequence, but rather through genetic linkage.

Topic Test 2: Relationship Between Gene Sequence and Protein Sequence

True/False

1. A single-base-pair change can alter protein activity enough to change phenotype.
2. Collinearity means the activity of the protein is determined by its gene.

Multiple Choice

3. Yanofsky's evidence for the relationship between genes and proteins did not include
   a. valine replacing glutamic acid in position 6 of the protein.
   b. alterations in the *trpA* gene.

c. that mutations in 5′ part of the gene affect the N-terminal end of the protein.
d. that mutations in 3′ part of the gene affect the C-terminal end of the protein.
e. that mutations in the central part of the gene affect the central part of the protein.

4. Which of the following is not true of Ingram's study?
   a. He studied hemoglobin.
   b. He used fragments of the protein.
   c. He studied normal versus abnormal proteins.
   d. He correlated the amino acid difference with a single-base-pair difference.
   e. He found a single amino acid difference.

Short Answer

5. Discuss Ingram's and Yanofsky's results in light of the central dogma (which was proposed at about this time).

Topic Test 2: Answers

1. **True.** Sickle-cell anemia is the example of this concept in this topic. This terrible disease is caused by a single-base-pair mutation.
2. **False.** Collinearity means the sequence of the gene directly determines the sequence of the protein.
3. **a.** Valine replaces glutamic acid in position 6 of the sickle-cell allele of hemoglobin. This was shown by Ingram.
4. **d.** Ingram did not know the nature of the DNA change in the sickle-cell allele.
5. The central dogma states that DNA is copied to RNA, which is used to make a protein. Ingram showed that heritable factors (which were known to be DNA) caused specific alterations in amino acid sequence of a specific protein. Yanofsky showed that mutations in DNA were collinear with alterations in amino acid sequence. Both of these observations are consistent with the central dogma.

TOPIC 3: COMPLEMENTATION ANALYSIS

KEY POINTS

✓ *What is the functional definition of a gene?*

✓ *What is a complementation test?*

✓ *How is a complementation test performed?*

Genes are functionally defined by the complementation test. The simplest case of complementation was seen in Chapter 1 when Mendel's heterozygous F_1 peas had a wild-type phenotype. The wild-type (round) allele of the seed-shape gene complemented the wrinkled allele, relieving its defect. In practice, the complementation test is used to define genes when two or more independent mutations produce the same or nearly the same phenotypes. Complementation testing is the simplest way to determine whether all mutations map to the same gene or each to its own gene or a combination of these two options. When Beadle and Tatum began their study of auxotrophic mutants in *Neurospora*, they created many mutants with the same phenotype, that is, the inability to grow in the absence of arginine. These mutants could have been mutant for the same single gene. Alternatively, the mutations could lie in separate genes, with those gene products forming a biosynthetic pathway to make arginine. The complementation test differentiates between these two explanations by sorting the mutants into groups, each of which identifies a gene.

The test is performed by making heterozygotes from true-breeding strains of each recessive mutation that has the common phenotype. Every pair-wise combination must be created and the phenotype of the heterozygotes must be noted. If the heterozygote produced by one of these crosses is wild type, then the mutations lie in two different genes, and the mutations are said to **complement**. If the heterozygote has the mutant phenotype, the mutations lie in the same gene, and they "fail to complement." This latter heterozygote is a particular kind of heterozygote: one that has two, differently mutant alleles of the same gene, said to be **heteroallelic**.

One important caveat to the complementation test is that it must be performed with recessive mutations only. Dominant mutations are not useful for complementation tests because dominant alleles cause mutant phenotypes even if the other allele of the gene is wild type. Hence, dominant mutations can be recognized in complementation tests by their failure to complement any other mutation. A second possible explanation for this behavior in complementation tests is that the other mutations map to adjacent genes and the universally noncomplementing mutation is a deletion of the contiguous genes. Other forms of analysis (such as mapping) would be necessary to distinguish between these possibilities.

A complementation test is a test of gene function. If the two mutations being tested in one complementation test lie in the same gene, there will be no functional gene product in the heterozygote and it will have the mutant phenotype. Only if the mutations lie in different genes will there be a wild-type allele of each gene also present in the heterozygotes, resulting in a wild-type phenotype. In this case, the mutations identify separate functions, usually separate genes.

Topic Test 3: Complementation Analysis

True/False

1. Only recessive mutations can be analyzed in complementation tests.
2. Complementation tests are used to identify mutations that are alleles of the same gene.

Multiple Choice

3. A researcher interested in spine development isolated four different strains of mice, all of whom have coiled tails. Each strain is true breeding. To study these mutations further, the researcher crossed each strain in every pair-wise combination and noted the curvature of the tail in their progeny. The results are shown below where "+" indicates straight tails and "–" indicates coiled tails.

	1	2	3	4
1	–	–	+	+
2	–	–	+	+
3	+	+	–	–
4	+	+	–	–

How many genes are represented by these strains?
a. None
b. One
c. Two
d. Three
e. Four

4. How are complementation tests executed?
   a. All mutants that have the same phenotype are crossed to make all possible heterozygotes.
   b. All recessive mutants that have the same phenotype are crossed to make all possible heterozygotes.
   c. All dominant mutants that have the same phenotype are crossed to make all possible heterozygotes.
   d. All mutants that have the same phenotype are crossed to get double homozygotes.
   e. All recessive mutants that have the same phenotype are crossed to get double homozygotes.

Short Answer

5. The fictional organism phamluster is red. Mutants are often green, pink, or blue. Two green phamlusters mated and had 12 offspring, all of who were red. Explain these offspring and give genotypes.

6. Haploid yeast (*Saccharomyces cerevisiae*) were mutagenized and seven independent isolates of histidine auxotrophs were identified. To determine which mutations lay in the same gene(s), complementation analysis was performed by mating together the haploid strains and testing the diploid for auxotrophy. The results are shown below where "+" means the diploid was able to grow on minimal medium without added histidine and "–" means the diploid required histidine supplement for growth.

	1	2	3	4	5	6	7
1	–	+	–	+	–	+	+
2	+	–	+	+	+	+	–
3	–	+	–	+	–	+	+
4	+	+	+	–	+	–	+
5	–	+	–	+	–	+	+
6	+	+	+	–	+	–	+
7	+	–	+	+	+	+	–

How many genes are identified by these mutations and which mutations are in the same gene(s)?

Topic Test 3: Answers

1. **True.** Dominant mutations fail to complement anything, so they are not interpretable.
2. **True.** If mutations map to the same gene, then they will not complement in a heterozygote. Alternatively, if the heterozygote has a wild-type phenotype, the mutations complement and identify different genes that are involved in the same cellular process.
3. **c.** One gene is identified by the mutations in strains 1 and 2, because these two mutations do not complement each other but do complement the other mutations. A second gene is identified by the mutations in strains 3 and 4, because these two mutations do not complement each other but do complement the other mutations.
4. **b.** Dominant mutations cannot be used. The informative progeny are hybrids whose parents were true breeders. These progeny are either heteroallelic for a single gene and have the mutant phenotype, or heterozygous for two genes (they have one wild-type allele of each) and have wild-type phenotypes.
5. The parents' "green" mutations are complemented in their offspring, who exhibit the wild-type phenotype. At least two genes are needed to convert green to red in this organism. Let one be *R* for the wild type and *r* for the mutant. Let one be *S* for the wild type and *s* for the mutant. The parents were *rrSS* and *RRss*; their progeny are *RrSs*.
6. There appear to be three genes represented by these seven mutations. Mutations 1, 3, and 5 identify one gene. Mutations 2 and 7 identify a second gene. Mutations 4 and 6 identify a third gene.

IN THE CLINIC

Many human diseases have a genetic basis. Typically, a mutation in DNA alters the amino acid sequence of the encoded enzyme, thereby altering the activity of that enzyme. One example is the autosomal recessive disease phenylketonuria (PKU). In normal individuals, excess phenylalanine (an amino acid obtained from protein in the diet) is converted to tyrosine, another amino acid and a precursor for other important molecules. Tyrosine can also be degraded through the pathway Garrod detected by its absence in alkaptonurics. Among the molecules made from tyrosine is dopa, a precursor to several important compounds including melanins and neurotransmitters. Phenylketonuric patients cannot convert excess phenylalanine to tyrosine, usually because they make mutant form(s) of the enzyme phenylalanine hydroxylase. Instead, the phenylalanine is degraded by an alternate pathway that creates substances toxic to developing nerve cells. If PKU babies are not treated, accumulation of the alternate breakdown products leads to severe mental retardation. Fortunately, there is an effective (usually) treatment for PKU. Begun within the first 2 months after birth and continued until the mid-teens, a restrictive diet that contains very little phenylalanine prevents the toxic by-products of faulty metabolism from accumulating, thereby preventing brain damage. Women who have PKU must resume the diet during pregnancy so that their offspring will not be subjected to the high levels of toxic by-products in their mother's bloodstream and will develop normally. By state law, newborns are screened for PKU, which affects about 1 of every 10,000 newborns in the

United States. Screening for PKU in newborns is humanitarian, and justified for this reason alone. In addition, testing all newborns costs one-tenth what it would cost to institutionalize all affected homozygotes. Incidentally, the failure to convert tyrosine to dopa by mutant enzyme tyrosinase results in albinism. Note that marriage between an albino person and a person with PKU would result in wild-type offspring, because the parents' mutations are in different genes and so would be complemented in their offspring.

DEMONSTRATION PROBLEM

Question: A researcher who wishes to study the biosynthesis of aromatic amino acids in yeast collects four haploid mutant strains, each of which will not grow on minimal medium unless it is supplemented with the amino acid tyrosine. Results of the complementation analysis the researcher performed are displayed in the chart below, where "–" means the diploid did not grow on minimal medium and "+" means the diploid grew on minimal medium. (NOTE: The chart is shown in an alternative manner that does not duplicate entries. All of the information that is needed to answer the question is shown.)

Mutants 1 and 3 were selected for further study. Each was tested for growth on minimal medium that had been supplemented with one of the known precursors to tyrosine, *p*-hydroxyphenylpyruvate or prephenate. Mutant 1 was able to grow in the presence of *p*-hydroxyphenylpyruvate, but mutant 3 could not. Neither strain could grow in the presence of prephenate. How many genes are identified by the four mutant strains? Diagram the relationship between these metabolites and gene(s).

	1	2	3	4
1	–	–	+	–
2		–	+	–
3			–	–
4				–

Answer: Mutations 1 and 2 map to the same gene (*A*), as shown by their failure to complement each other in the diploid. Mutation 3 maps to a second gene (*B*), as shown by its ability to complement mutations 1 and 2. Mutation 4 appears to be dominant, since it complements nothing. We cannot say whether it maps to gene *A*, gene *B*, or a third gene. Alternatively, genes *A* and *B* may map close together and mutation 4 is a deletion of both. It is not possible to distinguish between these explanations from these data.

Mutation 1 (gene *A*) can grow on more intermediates than can mutation 3 (gene *B*), suggesting it occurs earlier in the pathway than gene *B*. Therefore the synthetic pathway must be

$$\text{prephenate} \xrightarrow{\text{gene } A} p\text{-hydroxyphenylpyruvate} \xrightarrow{\text{gene } B} \text{tyrosine}$$

This pathway allows gene *A* mutants to be rescued by both *p*-hydroxyphenylpyruvate and tyrosine (all compounds that lie downstream of the inactive enzyme), while gene *B* mutants can only be rescued by tyrosine (because gene *B* mutants are blocked at this later step).

Chapter Test

True/False

1. Dumpy ("*dpy*") mutants in *Caenor elegans* have short, wide bodies. Two true-breeding lines of *dpy* animals were crossed and the hybrid progeny were wild type. These two lines have mutations in the same gene.
2. Garrod showed recessive inheritance of alkaptonuria.
3. A mutant phenotype results from alteration of several base pairs of a gene.
4. Complementation testing requires double homozygous recessive individuals.
5. Ingram's study of sickle-cell hemoglobin demonstrated specific effects of heritable mutations on proteins.
6. The interpretation of the *arg* mutants' inability to grow without arginine is that *arg* genes determine the ability to metabolize sugars.
7. Yanofsky showed collinearity for genes and their protein products.
8. Complementation tests define genes by function.

Multiple Choice

9. The biosynthetic pathway diagrammed below is responsible for feet color in an obscure bird. An orange-footed individual is crossed to a yellow-footed individual. What color are their F_1 generation feet? (Assume mutations are recessive.)

$$\text{white} \xrightarrow{enzA} \text{yellow} \xrightarrow{enzB} \text{orange} \xrightarrow{enzC} \text{brown}$$

   a. White
   b. Yellow
   c. Orange
   d. Brown
   e. Yellow and orange

10. If the F_1 individuals in the previous question are crossed, what is the ratio for the color of the F_2 generation feet?
   a. 9 brown:3 orange:3 yellow:1 white
   b. 9 brown:3 orange:4 yellow
   c. 9 brown:7 orange
   d. 12 brown:3 orange:1 yellow
   e. 15 brown:1 orange

11. Consider the pathway in question 9. What is the ratio for the F_2 generation from a cross of a white individual to a yellow individual?
   a. 9 brown:3 orange:3 yellow:1 white
   b. 9 brown:3 yellow:4 white
   c. 9 brown:7 white
   d. 12 brown:3 yellow:1 white
   e. 15 brown:1 white

12. Which of the following was not involved with the proof of the gene-protein relationship?
    a. Charles Yanofsky
    b. Vernon Ingram
    c. Collinearity of DNA mutation and altered amino acid
    d. Sickle-cell anemia mutation
    e. Cystic fibrosis mutation

13. The primary function of complementation testing is
    a. to identify dominant alleles.
    b. to identify recessive alleles.
    c. to identify the number of genes represented by a set of mutations.
    d. to determine metabolic pathways.
    e. to accentuate mutant phenotypes.

14. Two recessive alleles ("*1*" and "*2*") give the same phenotype when homozygous. These homozygotes are crossed and their progeny have the wild-type phenotype. What explains this result?
    a. Compensation
    b. Complementation
    c. Consideration
    d. Hybridization
    e. Recombination

Short Answer

15. The pathway below is from a question in Topic 1 involving black or blue fur in a mysterious organism. Assume that mutations in these genes are recessive. What would be the color of the offspring if a blue individual were crossed with a white individual? Explain your answer.

$$\text{colorless} \xrightarrow{enzA} \text{white} \xrightarrow{enzB} \text{blue} \xrightarrow{enzC} \text{black}$$

16. A typical dihybrid F_2 ratio that indicates gene interaction is 9:3:4 (recessive epistasis). You may recall from Chapter 3 that coat color in Labrador retrievers is determined by two genes. The *B* gene causes a black pigment to be made; brown is recessive. The *E* gene causes the pigment to be deposited in the hair fiber; the recessive allele results in no pigment, a golden color. Draw a pathway for these two genes, depicting their order of action.

17. Describe the reasoning behind the ordering of Beadle and Tatum's *arg* mutants.

Essay

18. Fred, a geneticist, isolated seven auxotrophic mutants of yeast who cannot make the steroid ergosterol. All were recessive and when Fred crossed them together in all combinations, he discovered that some of the heterozygotes were no longer auxotrophic. This result is shown in the chart as "+," while auxotrophy is represented by "–."

	1	2	3	4	5	6	7
1	–	+	+	–	–	+	+
2		–	+	+	+	–	+
3			–	+	+	+	–
4				–	–	+	+
5					–	+	+
6						–	+
7							–

After discovering some related compounds, Fred tests mutants 3, 5, and 6 for their ability to grow on minimal medium that has been supplemented with one of these compounds. Mutant 3 grows when supplemented with ergosterol. Mutant 5 grows when supplemented with ergosterol and lanosterol and zymosterol. Mutant 6 grows when supplemented with ergosterol and zymosterol. None of the mutants will grow when squalene is the supplement, even though squalene is a known precursor of ergosterol. From the data provided, discuss a likely pathway for the synthesis of ergosterol and reasonable positions for all seven mutants in that pathway.

Chapter Test Answers

1. **F** 2. **T** 3. **F** 4. **F** 5. **T** 6. **F** 7. **T** 8. **T** 9. **d** 10. **b** 11. **b** 12. **e** 13. **c** 14. **b**

15. The blue individual is homozygous for *enzC* mutant alleles and the white individual is homozygous for *enzB* mutant alleles. Their offspring will be heterozygous for both genes. Because they will have one wild-type allele of each gene, they will have wild-type phenotypes; that is, their mutant alleles are complemented.

16. brown $\xrightarrow{\textit{gene B}}$ black $\xrightarrow{\textit{gene E}}$ hair

17. Mutants that grew on few compounds were blocked for later steps in the pathway. Mutants that grew on a greater number of compounds were blocked for earlier steps in the pathway.

18. The seven mutations identify three genes that are important for ergosterol biosynthesis. Mutants 1, 4, and 5 identify the same gene. Mutants 3 and 7 identify a second gene. Mutants 2 and 6 identify a third gene. Hence, each of the mutants that were tested further represent different genes. The mutant that grows on the fewest number of compounds acts the latest in the pathway: Mutant 3 only grows in the presence of ergosterol, so it is likely the last of the three genes to act in the wild type. Likewise, mutant 6 grows on fewer compounds than mutant 5, so 6 occurs later than 5. The best pathway to represent these genes is

$$\text{squalene} \xrightarrow{\textit{mut5}} \text{lanosterol} \xrightarrow{\textit{mut6}} \text{zymosterol} \xrightarrow{\textit{mut3}} \text{ergosterol}$$

Check Your Performance:

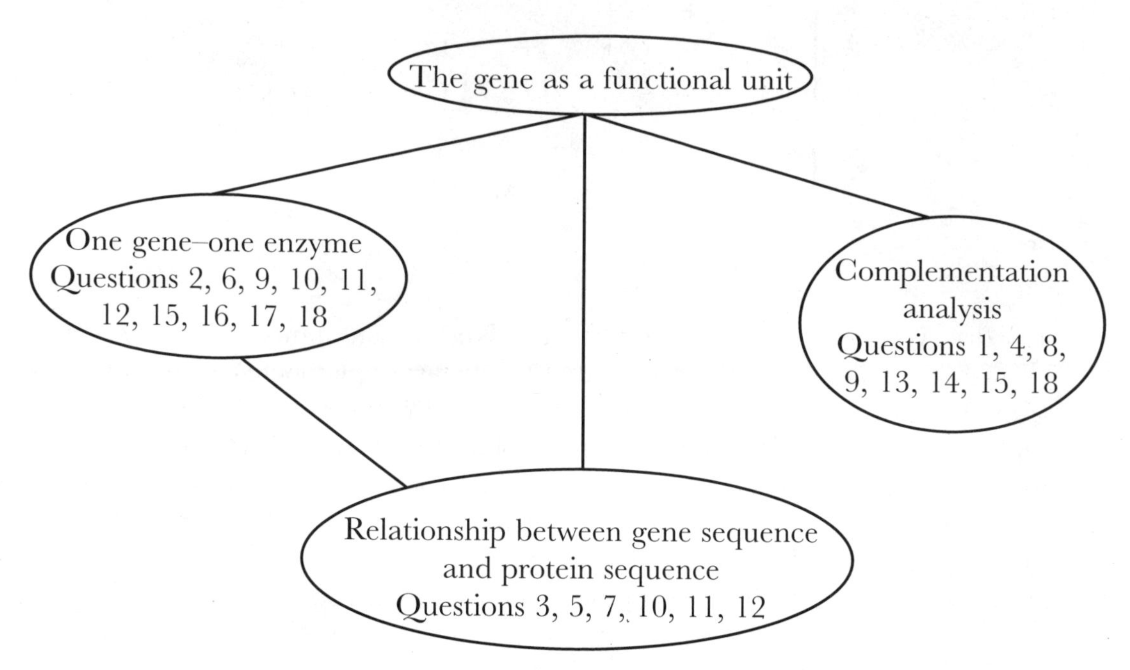

Use this chart to identify weak areas, based on the questions you answered incorrectly in the Chapter Test.

Techniques of Molecular Genetics

Genetics has blossomed into a new arena since the mid 1970s when it became possible to manipulate fragments of DNA in bacteria. Molecular understanding of genes and molecular approaches to understanding gene function have broadened the knowledge of genetic phenomena. It is now possible to predict whether unborn children carry one of several deadly genetic diseases, and to predict one's own fate for several more. This knowledge creates many ethical dilemmas for society that an informed citizenry will have to decide how best to manage. This chapter focuses on several of the more common techniques for investigating the molecular nature of heredity and, increasingly, for diagnostics.

ESSENTIAL BACKGROUND

- **Pedigree analysis (Chapter 3)**
- **Sex linkage (Chapter 3)**
- **Linkage (Chapter 4)**
- **DNA structure and replication, mRNA, and gene expression (Chapter 5)**
- **Null mutations (Chapter 6)**
- **Complementation (Chapter 8)**

TOPIC 1: RECOMBINANT DNA

KEY POINTS

✓ *What is recombinant DNA?*

✓ *What are vectors?*

✓ *What are the essential features of vectors?*

In the mid 1970s, a class of enzymes discovered in bacteria proved to have exceedingly useful properties for researchers. These **restriction endonucleases** cleave the DNA duplex at specific sites. Each restriction enzyme recognizes a specific nucleotide sequence, usually four or six nucleotides long and having symmetry such that the sequence reads the same on both strands. For example, the enzyme *Eco*RI recognizes and cleaves the sequence shown (below at left) wherever it occurs in the DNA. Notice that the upper strand is the same sequence in the 5′ to 3′ direction as is the lower strand in its 5′ to 3′ direction. Because of this symmetry (and despite the polarity of the strands) restriction sites are sometimes said to be **palindromes**. Most enzymes cause symmetrical, but staggered cleavage of the DNA backbone at their recognition sites, producing fragments that have single-stranded tails (**cohesive** or **sticky ends**) of one to four nucleotides (*Eco*RI leaves 4-bp overhangs, shown at right). Some enzymes cleave the backbones directly opposite each other, producing fragments without cohesive ends, called **blunt ends**.

5′-GAATTC-3′ ⟶ 5′-G AATTC-3′
3′-CTTAAG-5′ 3′-CTTAA G-5′

The number of fragments produced by restriction enzyme digestion depends on the number of recognition sites within and the form of the DNA. Digestion of circular DNA produces one fragment per site. Digestion of linear DNA produces one more fragment than there are sites. Any DNA digested with a given enzyme will have the same cohesive ends (i.e., single-stranded overhangs). DNA from a mammalian cell that has been digested with *Eco*RI has the same sticky ends as DNA from a bacterial cell that has also been digested with *Eco*RI. If digested DNA from the two species is mixed together, the complementary ends pair with each other and the enzyme **DNA ligase** can form new covalent bonds between the paired ends, creating novel junctions between the mammalian and bacterial DNA. **Recombinant DNA** is any new DNA sequence formed by combining DNA from two or more sources. Often, this consists of foreign DNA that has been spliced into a small replicating DNA molecule (often by virtue of restriction sites). This recombinant DNA molecule can be copied, or **cloned**, many times by a host cell.

The foreign DNA can be from any source; two general types are discussed in Topic 2. The small replicating DNAs, called **vectors**, have three important features. First, they must have at least one restriction site at which foreign DNA can be introduced. Second, they must be able to be amplified by at least one organism, usually bacteria such as *Escherichia coli*. Third, vectors must be selectable, meaning that the researcher must be able to grow the cells that contain the vector to the exclusion of the cells that lack it. Selection can take several forms; the most common is a gene (a **selectable marker**) that confers resistance to an antibiotic such as ampicillin (*amp*) or tetracycline (*tet*), and thus is dominant. Bacteria that contain the vector can grow and form colonies on agar plates that contain the antibiotic. Those that lack the vector will not form colonies because they are sensitive to the antibiotic. Some plasmids contain two different antibiotic-resistance genes (e.g., both amp^R and tet^R), so that cloning a foreign DNA into one gene disrupts its function while leaving the other intact. In this way, the background of vectors that reclose without insert can be eliminated. Bacteria that are sensitive to the appropriate antibiotic but resistant to the other carry insert-containing vectors. Several types of vectors meet all three requirements. The oldest type is the **plasmid**. Plasmids grow as autonomously replicating circular DNAs in bacteria and are useful for cloning foreign DNA **inserts** ranging up to 20 kb in size. The first widely used plasmid was pBR322, which is diagrammed in **Figure 9.1**. Many of today's vectors are derived from this plasmid.

A second type of vector is based on the bacteriophage **lambda (λ).** Wild-type phage grows by infecting a bacterial host, replicating and packaging new virus particles, and lysing the cell to get out (the lytic cycle). The lysis is visualized as a clear **plaque** (of lysed cells) on an opaque layer of uninfected host cells. The central portion of the viral genome contains genes that are dispensable for the lytic cycle. This region can be removed by digestion with restriction enzymes and foreign DNA inserted can be in its place. The advantage of using these vectors is that a restriction on genome size causes only vectors that contain inserts to be packaged into infective phage particles. Consequently, only phages that contain inserts will be produced. A disadvantage is that lambda can only carry inserts of 10 to 15 kb.

Cosmids are vectors that are hybrids between plasmids and phages. They can be packaged like phage and therefore can be size selected for inserts ranging from 35 to 45 kb. Yet, they can be

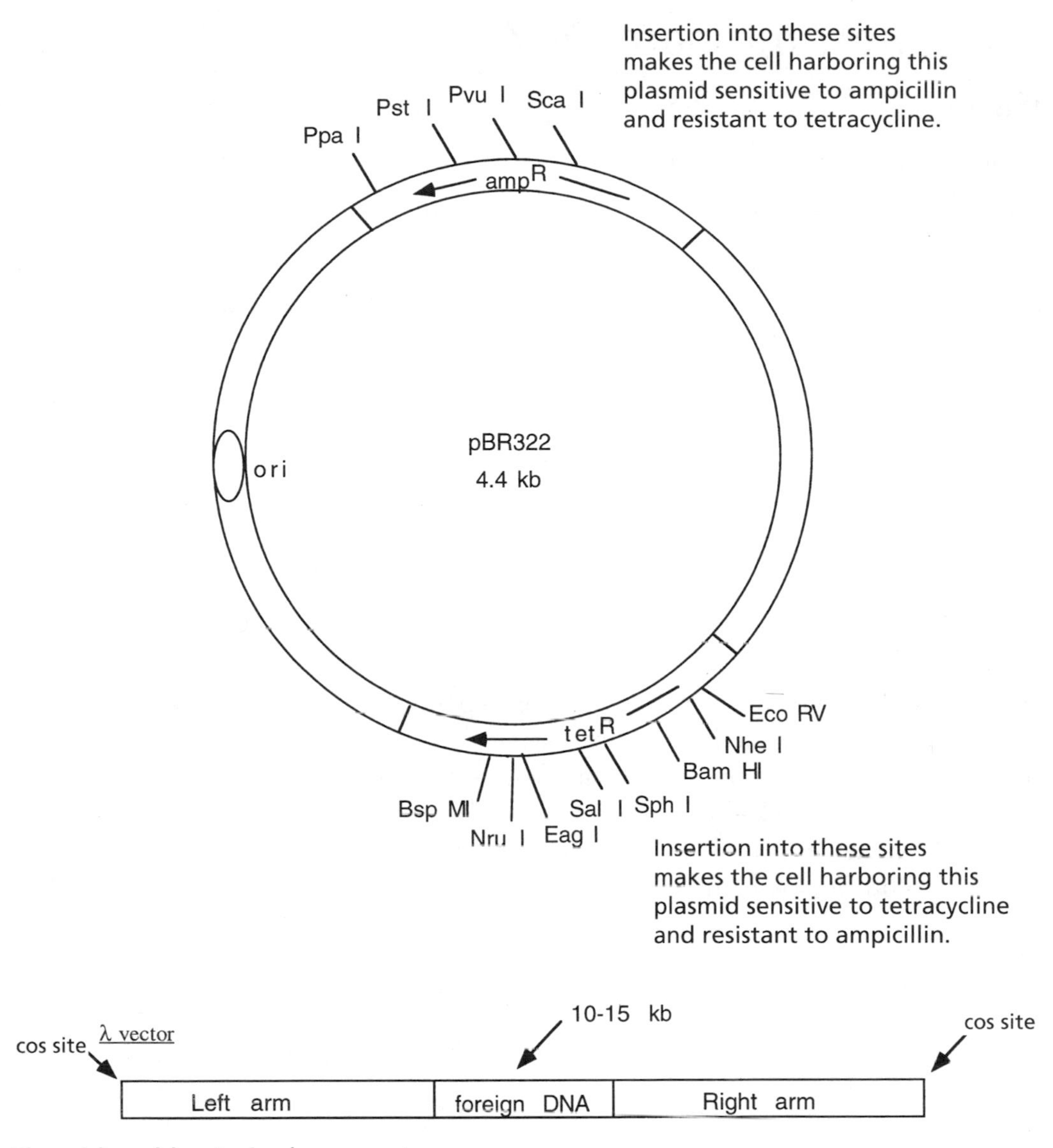

Figure 9.1 Plasmid and bacteriophage vectors.

grown like a plasmid, by allowing colonies to form on selective medium. Essentially, cosmids are plasmids to which the signal sequences for phage head packaging (**cos sites**) have been added. Cosmids do not have genes for the lytic cycle, so they can contain larger pieces of cloned DNA.

Some vectors are designed to grow in more than one species. **YACs**, or yeast artificial chromosomes, are vectors that act like plasmids in bacteria and like chromosomes in yeast. They are selectable in both species and can be used to shuttle genes between the two species for genetic studies of gene function. When grown in yeast, YACs can handle inserts of several hundred kilobases, making them the preferred vector for work with the larger genes characteristic of mammals. Other vectors grow in plant, insect, or mammalian cells and these usually grow in bacteria, as well. Nearly all vectors have been designed to grow in bacteria because this is a fast, comparatively inexpensive means of isolating large quantities of an interesting sequence.

Topic Test 1: Recombinant DNA

True/False

1. Cohesive enzymes are enzymes that cleave DNA at specific sequences.
2. Clones are only exact genetic replicas of organisms.
3. If a linear DNA has three recognition sites for a particular restriction endonuclease, digestion with this restriction endonuclease creates three fragments.

Multiple Choice

4. Which of the following is not a vector?
   a. Plasmid
   b. Lambda
   c. Cos
   d. Cosmid
   e. YAC
5. Which of the following is not a selection scheme for vectors?
   a. Ampicillin resistance in cells harboring the plasmid
   b. Tetracycline resistance in cells harboring the plasmid
   c. Right size for packaging lambda phage DNA
   d. Tetracycline resistance in cells infected with lambda phage
   e. Ampicillin resistance in bacteria harboring YACs

Short Answer

6. Distinguish between the three major types of bacterial vectors.

Topic Test 1: Answers

1. **False.** These enzymes are called *restriction endonucleases*. Cohesive ends are complementary.
2. **False.** Clones are exact replicas of DNA (fragments or whole organisms' genomes). The common understanding of *clone* is also a valid meaning.
3. **False.** The situation described will produce four fragments.
4. **c.** The cos sites are the packaging signals for the lambda bacteriophage.
5. **d.** Lambda does not contain antibiotic-resistance genes. Selection of recombinant phage is based on the size of the genome. The recombinant phage must be about 50 kb.
6. The three major types are plasmids, phages, and cosmids. Plasmids grow as small, circular, self-replicating DNAs in bacteria. They accommodate inserts up to 20 kb. Phages grow by infecting and lysing bacteria. Infectious particles contain inserts of 10 to 15 kb. Cosmids replicate like plasmids but can be packaged into phage particles like lambda. They accommodate inserts of 35 to 45 kb. Bacteria that carry plasmids and cosmids are selected by resistance to a specific antibiotic. Phages are selected according to size: Only those that are the right size to be packaged into the particle will be infectious.

TOPIC 2: GENOMIC AND cDNA LIBRARIES

KEY POINTS

✓ *What is a library?*

✓ *How are libraries made?*

✓ *How are libraries used?*

One application of recombinant DNA is the creation of libraries. A **genomic library** is a collection of fragments of DNA encompassing the entire genome of an organism. Genomic libraries are made by digesting total cellular DNA with a restriction enzyme and ligating the resulting fragments to vector molecules. This creates thousands to millions of different plasmids, a complexity that is useful for identifying multiple genes, as discussed later. This library is maintained as a mixed collection of clones. It can be used to isolate interesting genes from a particular organism or to study properties of that organism's genome.

It is also possible to make libraries from messenger RNA (mRNA). Bigger eukaryotes (e.g., vertebrates) have significant amounts of noncoding DNA in their genomes. Genomic libraries, therefore, contain many clones that represent less-interesting DNA. Hence, screening such libraries for a particular gene entails looking through a lot of "junk." One way to eliminate much of the junk is to make a library of transcribed sequences (total RNA). Since most cellular RNA does not encode protein, further refinement of the library can be achieved by making it solely from mRNA, which is only 1% to 10% of total cellular RNA.

Because differing cell types in higher eukaryotes express different genes (see Chapter 5), expression libraries are cell-type specific, as well as organism specific. Restriction enzymes do not cleave mRNA, so cannot be used to clone these molecules. Library construction is accomplished with **reverse transcriptase**, an RNA-dependent DNA polymerase first isolated from RNA-based viruses (**retroviruses**). This enzyme uses RNA templates to synthesize a DNA strand. The RNA strand is removed by **RNase H**, an enzyme that degrades the RNA strand in an RNA-DNA hybrid. While the degradation is occurring, short incompletely digested RNAs are present and can be used by **DNA polymerase I** as primers for the synthesis of the second DNA strand. The resulting double-stranded DNA has blunt ends (i.e., no overhangs). It is cloned into vectors and maintained as a complex collection of different sequences. Because these synthesized DNAs are complementary to the mRNAs, this library is called a **complementary DNA (cDNA) library**.

Libraries are used to identify specific genes of interest. If a fragment of the gene has already been cloned, it can be used to identify clones that contain more complete sequences of the gene. (Often individual clones within cDNA libraries do not contain the complete gene sequence.) Or if the gene product is known, the library can be screened for its activity or for a diagnostic DNA fragment. For the purpose of identifying a specific gene, possibly the most common use of a genomic library is to transform (i.e., introduce) it into bacterial or yeast cells that contain null mutations in that gene. Rarely, one of the transformants, identifiable by expression of the selectable marker, will have the wild-type phenotype instead of the mutant null phenotype. The defect in such a transformant is complemented by a wild-type allele of that gene from the library. In this way, the gene of interest may be identified definitively.

Topic Test 2: Genomic and cDNA Libraries

True/False

1. The cDNA molecules are made from all of the RNA within a cell.
2. Complementation can be used to identify library clones of interesting genes.

Multiple Choice

3. To create a cDNA, all of the following items are needed except
   a. mRNA.
   b. RNase H.
   c. DNA polymerase I.
   d. restriction enzymes.
   e. reverse transcriptase.
4. Which of the following enzymes synthesizes DNA from an RNA template?
   a. DNA polymerase I
   b. DNA ligase
   c. RNase H
   d. Reverse transcriptase
   e. Restriction enzyme
5. Which of the following statements is false?
   a. Genomic libraries contain coding and noncoding sequences.
   b. The cDNA libraries contain only coding sequences.
   c. Libraries are specific to organisms and represent all cell types.
   d. Libraries can be used to identify a clone of a particular gene.
   e. Libraries are random collections of assorted sequences.

Short Answer

6. To find a clone of the mouse gene encoding actin (a protein), would it be better to screen a genomic library or a cDNA library? Why?

Topic Test 2: Answers

1. **False.** The cDNA molecules are made from mRNA only.
2. **True.** The library must be constructed so that the gene can be expressed by the organism carrying the null mutation, but this is an established method of identifying genes.
3. **d.** These enzymes do not cleave mRNA or RNA-DNA hybrids and the double-stranded DNA produced does not necessarily have restriction sites. The mRNA is converted into double-stranded DNA, which is ligated into vectors via its blunt ends.
4. **d.** The name of this enzyme refers to the "backward" flow of information: from RNA into DNA. It is used by RNA-genome viruses to make DNA copies of their genome upon infection. It is also essential for making cDNAs because no other known enzyme can perform this function.

5. **c.** The cDNA libraries are cell-type specific as well as being organism specific.
6. A cDNA library would be better because it is enriched for protein-coding sequences so fewer clones would have to be screened.

TOPIC 3: RESTRICTION MAPPING AND RFLP MAPPING

KEY POINTS

✓ *What are RFLPs?*

✓ *What is RFLP mapping?*

✓ *What is restriction mapping?*

DNA has a negative charge, proportional to the length of the fragment, as a result of the phosphates in the backbone. Consequently, it is possible to separate the DNA fragments generated by restriction enzyme digestion on the basis of their size by their differential migration in an electric field termed size-fractionation. The digested DNA is loaded into one end of a horizontal agarose or vertical polyacrylamide gel (imagine very thick gelatin) and an electric current is applied to the tank containing the gel, with the positive terminal at the end of the gel opposite the DNA and the negative terminal at the end nearest the DNA. The negatively charged DNA migrates through the gel toward the positive pole, with the smallest fragments moving the fastest and the largest moving the slowest because they are more retarded by the gel. The fragments can be visualized with a fluorescent dye (usually ethidium bromide) that binds to DNA. If the digested DNA is from a clone, the fragments of the clone will appear as discrete "bands" of fluorescence. The bands reveal the sizes of DNA fragments when compared to a size marker included on the gel.

Digestion of the same plasmid with several different enzymes, singly and in pairs, will enable the researcher to create a **restriction map** of the plasmid and this facilitates manipulation of the insert. **Restriction mapping** is used to determine the relative positions of restriction sites in a cloned piece of DNA so that the DNA can be manipulated further or to identify neighboring (overlapping) fragments in other clones by overlapping maps. The assembly of the restriction map is a logical exercise in finding the best map that accounts for all digestion fragments.

In contrast, when genomic DNA is size-fractionated on a gel, the dye reveals a smear with no discrete bands, due to the complexity of the fragments. Random spacing of restriction sites generates randomly sized fragments. Digestion of total genomic DNA with a restriction enzyme generates a plethora of DNA fragments of diverse sizes. To visualize specific fragments from genomic DNA, the gel is "blotted" to a nylon or nitrocellulose membrane, to which the DNA will adhere. This immobilized DNA can be probed with another DNA fragment. Because complementary DNAs hybridize (**anneal**) to each other, the position of the fragment that is complementary to the DNA probe is revealed by the probe sticking to this portion of the blot. The probe is usually synthesized from radioactive phosphorus in one nucleotide. The blot can be exposed to x-ray film to obtain a scale image of the blot, with dark lines on the film indicating where it is in contact with the accumulated radioactive probe. Similar but nonradioactive methods can also be used to identify the location of the probe. This procedure for generating the film image of a specific DNA fragment within a genome is called **Southern blotting**, after its inventor E. M. Southern. The image is called a **Southern blot**.

The size of the fragment containing the probe depends on the location of the flanking restriction sites, which can vary. When the sites occur within coding regions, they tend to be the same in all individuals of a species. However, sites present in noncoding regions can vary among individuals. These latter sites are useful for locating genes on the physical map. One individual's DNA may contain restriction sites that give a particular fragment and another individual's DNA may be missing one of these sites or have a closer site, generating a differently sized fragment from the same DNA region. These differences are heritable and can be used as markers on chromosomes just like heterozygous genes are used (see Chapter 4). Thus, genetic maps can include these physical entities, as well as expressed genes. Differences in Southern blot patterns result from differences in the locations of restriction sites, giving this phenomenon the name **restriction fragment length polymorphisms**, or **RFLPs**. Particular RFLPs result from particular restriction enzyme and probe combinations; changing either of these components of the Southern blot produces a different blot, and defines a different RFLP. Individuals can be homozygous or heterozygous for the alleles of an RFLP, which can be named by size (e.g., 5 kb or 1 kb) or by arbitrary letter designations. RFLPs, like any other marker, can be mapped against other genes.

In the cloning of human disease genes, linked RFLPs are often known before the gene is identified. In the meantime, the RFLP can be used as a predictor for predisposition, if the RFLP-disease gene allele configuration in the family can be determined from each individual's affected relatives. The gene for a trait such as a heritable disease can be cloned if a RFLP is found to be tightly linked to it, indicating close physical distance between the RFLP and the trait's gene. In cases where complementation assay is possible, the RFLP serves as a starting point for testing DNA segments for their ability to complement the mutant gene (indicating the presence of the wild-type gene in the DNA fragment). The screen then works outward in both directions from the RFLP, testing overlapping clones for complementation. If no complementation test is possible, the outward search scans first for genes amid noncoding sequences. Once candidate genes have been identified, the coding sequences of these genes are compared in phenotypically wild-type and mutant individuals. The gene responsible for the trait being studied should differ in these two groups of individuals.

Topic Test 3: Restriction Mapping and RFLP Mapping

True/False

1. Individuals are always heterozygous for RFLP alleles.
2. The inheritance of RFLP alleles can be followed in parents and their offspring.
3. RFLP patterns are specific for a particular restriction enzyme and probe.

Multiple Choice

4. A person is heterozygous for the recessive allele that causes cystic fibrosis (cf^+/cf) and also heterozygous for a linked RFLP (*A1/A2*). If the allele configuration is $A1\ cf^+/A2\ cf$, what RFLP allele is likely to be found in offspring who inherit the disease allele?
   a. *A1*
   b. *A2*
   c. *A1* and *A2*
   d. Neither

5. If an RFLP is 3 m.u. from a disease gene, how often do you expect the RFLP allele to be "wrong" about whether the disease allele is present?
   a. 50%
   b. 25%
   c. 6%
   d. 3%
   e. 1.5%

Short Answer

6. A rare autosomal dominant disease is characterized by an absence of molar teeth. The RFLP data for a family that segregates the no-molars trait are shown below. DNA from individuals was obtained and analyzed by RFLP mapping to determine the location of the gene in which mutations prevent molars from forming. Affected individuals are indicated by filled symbols. The RFLP pattern for each individual lines up with that person's symbol. RFLP alleles are 2 kb, 3 kb, 5 kb, and 6 kb.

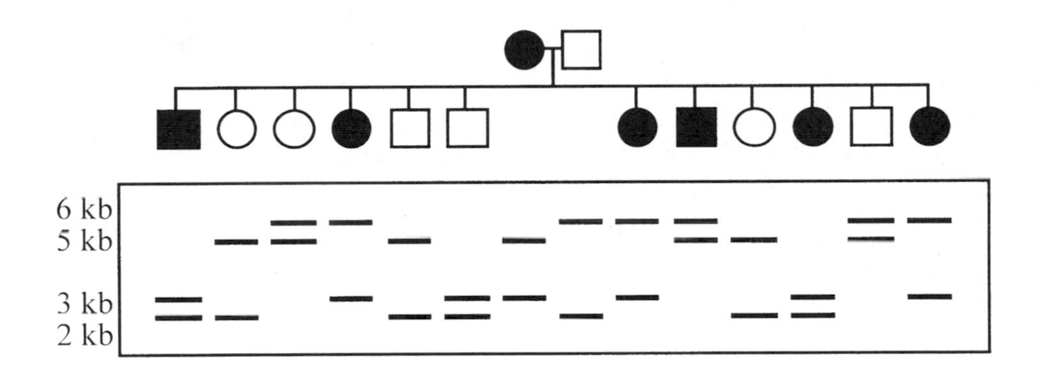

Which RFLP allele is *cis* to the no-molars gene in this family, or are the no-molars gene and the RFLP not linked? Explain the classes of progeny.

Topic Test 3: Answers

1. **False.** Like alleles of any coding gene, the RFLP alleles can be homozygous or heterozygous.
2. **True.** Pedigree analysis for genetic disease and RFLPs can identify linkage relationships between these two types of DNA sequence that facilitate study of the disease gene. This is possible because both types of sequences are heritable.
3. **True.** Changing either of these two parameters defines a new RFLP.
4. **b.** The *A2* allele is *cis* to the disease allele in the parent so most gametes that contain the disease allele will also contain the *A2* RFLP allele.
5. **d.** For 3 m.u. between the RFLP and the disease gene, 3% of the gametes in a dihybrid will be recombined for the interval. In these gametes, the previously *trans* RFLP allele is now *cis* to the disease allele and the previously *cis* RFLP allele is now *cis* to the wild-type allele of the disease gene.

6. The RFLP and no-molars gene appear to be linked. The mother's 3-kb allele is *cis* to the no-molars allele. Five of the progeny inherited the no-molars allele along with the 3-kb RFLP allele. Five of the progeny inherited the normal-teeth allele along with the 5-kb RFLP allele. Two of the progeny are recombinant for this interval, having the mother's 3-kb allele without the disease allele (II-6) or having the disease allele but not the mother's 3-kb allele (II-8). These exceptional progeny number fewer than 50% of all offspring, consistent with linkage of the RFLP to the no-molars gene (many, many more families of this sort would be needed to be certain of linkage).

TOPIC 4: SEQUENCING

KEY POINTS

✓ *How is the nucleotide sequence of DNA determined?*

✓ *What is a genome project?*

Once a cloned DNA fragment is identified as containing the gene of interest, the next goal is to know the sequence of nucleotides that compose the gene. There are two major techniques for sequencing DNA. The **Maxam–Gilbert** method is rarely used, but is the preferred method for determining the sequence of very small, uncloned DNAs. Four separate reactions are carried out, one for each nucleotide. In each reaction, one of four base-specific chemicals is used to cleave the DNA. The DNA is labeled on one end beforehand, so that cleaved molecules that contain the same end of the original DNA can be identified. The reactions are done in conditions that average one cut per molecule, no matter how many times the base appears. Thus, a mixed population of molecules is generated with lengths corresponding to the location of the specific base cleaved in that reaction. These lengths are measured on a polyacrylamide gel.

Sanger sequencing is also known as *dideoxy sequencing*, which describes one of the components. A dideoxyribonucleotide triphosphate (ddNTP) differs from a normal deoxyribonucleotide (dNTP) in that it does not have the 3′ hydroxyl to which a polymerase would attach the next nucleotide in a DNA synthesis reaction, thus causing chain termination. In Sanger sequencing, ddNTPs are randomly incorporated into the synthesized strand, in place of normal dNTPs. For example, dideoxythymidine triphosphate (ddTTP) causes accumulation of a population of molecules whose size varies but all of which end in thymidine. The researcher carries out four reactions with the cloned DNA, one for each ddNTP. The dideoxy sequencing reaction includes the DNA template, a DNA polymerase, a short DNA (oligonucleotide) primer, normal dNTPs, and one of the ddNTPs. Most vectors are arranged so that at least one common primer sequence is adjacent to the site where foreign DNAs are cloned. The primer is a synthetically manufactured DNA about 20 nucleotides long. In the sequencing reaction, it binds to the complementary site in the plasmid. Extension of the primer by DNA polymerase copies the insert.

To visualize the products, either the primer or one of the dNTPs is radioactively labeled. The length of each product relative to the primer identifies the nucleotide at that position in the chain. After synthesis, the products are electrophoresed on polyacrylamide gels to separate them by size and the gel is exposed to x-ray film. Sequence is read from the bottom of the gel (5′) to the top (3′), where a dark band on the gel represents a chain-termination event and therefore identifies the nucleotide at that position. The sequence read is the complement of the template strand.

A more recent innovation on this procedure greatly shortens the time needed to perform the reactions and read the sequence. This innovation uses polymerase chain reaction (PCR), a technology described in Topic 5. This technique is faster, simpler, safer, and less expensive than the radioactive method. All four dideoxy reactions are combined into a single reaction by using a different fluorescent dye as the label on each of the dideoxy terminators. A laser and computer read the gel, looking for each dye's characteristic fluorescent wavelength to identify each nucleotide by position, and creates data files of the sequence for direct analysis by the researcher. A major analysis involves comparing the DNA and predicted protein sequences to previously identified sequences in national and international databases. Finding similarities between the sequences of two gene products can suggest similarities in function or regulation.

Sequences of complete genomes are known for a number of model genetic organisms and clinically important organisms. **Genome projects** have two additional goals, which should now be familiar concepts. First, a physical map is created, noting the positions of RFLPs on a detailed restriction map. This map will direct the assembly of sequence segments (**contigs**) into their correct positions in the data set. Second, a genetic map is created from the knowledge of the physical map plus the genes that have already been identified. The Human Genome Project is the effort to sequence the human genome; it is expected to finish in 2003. Unfortunately, knowing the sequence of any organism's DNA is only the beginning to answering questions such as what is the best disease treatment.

Topic Test 4: Sequencing

True/False

1. All sequencing methods rely on size fractionation of the reaction products to determine sequence.
2. The sequence read from an x-ray film is exactly the sequence of the template strand in dideoxy sequencing.
3. One advantage to fluorescence-based sequencing is that there is one reaction per template instead of four.

Multiple Choice

4. What method is best used to sequence very small uncloned DNAs?
   a. Maxam–Gilbert method
   b. Sanger method
   c. Dideoxy sequencing
   d. Fluorescence-based sequencing
5. Which method of labeling *cannot* be used to detect the products of sequencing reactions?
   a. Labeled primer
   b. One labeled end
   c. Fluorescent dideoxyribonucleotides
   d. Radioactive nucleotides
   e. Labeled template

6. Which one of the following is not a component of a dideoxy sequencing reaction?
   a. Primer
   b. Template
   c. RNA polymerase
   d. Dideoxyribonucleotides
   e. Deoxyribonucleotides

Short Answer

7. How does chain termination in dideoxy sequencing enable sequencing?

Topic Test 4: Answers

1. **True.** The sequencing methods identify nucleotides by breaking the DNA backbone or by terminating the synthesis of the new DNA strand in a nucleotide-specific manner. Size fractionation of these molecules reveals the sizes of all molecules ending with the same nucleotide per reaction. Composite analysis of all four reactions per template yields the sequence by identifying the nucleotide at each position in the DNA strand.
2. **False.** It is the complement of the template strand.
3. **True.** This experimental design greatly reduces the handling time and the possibility for error in each reaction, thereby increasing throughput and accuracy.
4. **a.** Options b, c, and d require that the DNA to be sequenced be copied by a DNA polymerase, which is not possible for very small uncloned DNAs. Options b and c are two names for the same method. Option d is a variation on the method in b and c.
5. **e.** Labeling the template in the synthesis reaction will not enable visualization of size differences in the newly synthesized strands.
6. **c.** DNA polymerase is needed, not RNA polymerase.
7. In dideoxy sequencing, DNA polymerase copies the DNA whose sequence is desired into many new strands. These strands are frequently and randomly terminated by incorporation of a ddNTP. Because no nucleotides can be added after a dideoxyribonucleotide is incorporated, the collection of terminated strands indicates the positions of the ddNTP, and therefore also the normal analog, in the DNA template. Four separate reactions, one for each nucleotide, are carried out to provide a complete sequence for the DNA template used.

TOPIC 5: POLYMERASE CHAIN REACTION AND SEQUENCE TAGGED SITES

KEY POINTS

✓ *What is PCR?*

✓ *What is an STS?*

The realization that DNA polymerase could be used to make copies of a DNA template in vitro (i.e., outside a living organism) led to the invention of the **polymerase chain reaction (PCR)**. The concept is fairly simple: Two primers are generated with specific sequences. Each of the primers will hybridize to opposite strands of a template with their 3′ ends pointed toward each other. Extension of these primers by DNA polymerase will copy the intervening template, including the other primer's binding site (**Figure 9.2**). After the primers are extended, the reaction is heated to denature the newly synthesized strands from their templates, and then rapidly cooled so that the strands do not reanneal. Instead, the strands are bound by more primers (PCR is carried out with excess primers) and the polymerase extends these primers, as it did before. The reaction is cycled between denaturation and polymerization conditions repeatedly, theoretically doubling the number of templates in each round (geometric accumulation, see

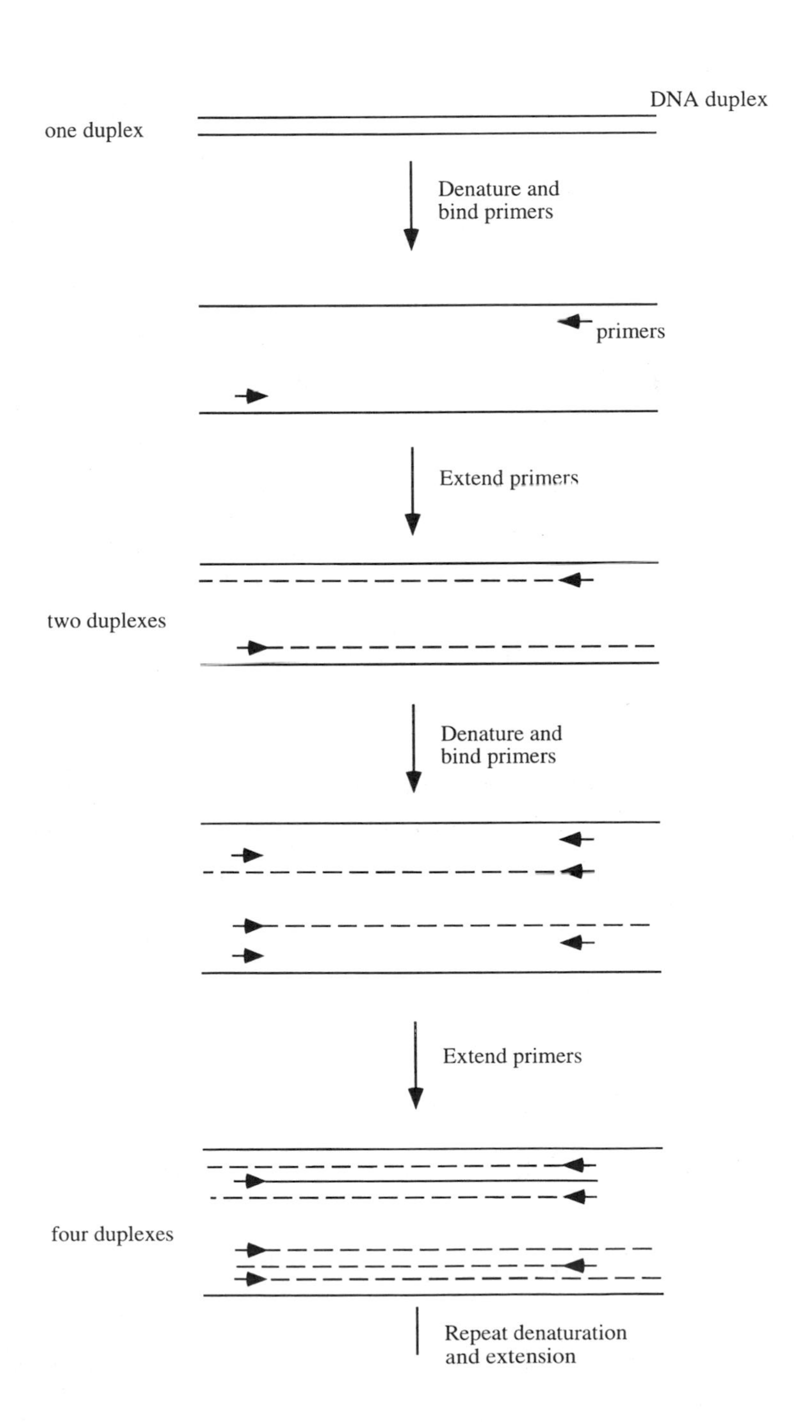

Figure 9.2 Polymerase chain reaction (PCR).

Figure 9.2). DNA from a single cell can yield millions of fragments of the sequence bounded by the primers after only 20 cycles. All that is required is that enough sequence of the template be known so that primers can be made. If no flanking sequence is known, PCR is not possible. Note that only the DNA between the two primers is amplified, and that the size of that amplified fragment is defined by the distance (on the template) between the primers. Hence, the product of PCR is a discrete size.

The discovery of thermostable DNA polymerases in bacteria that live in hot tubs and deep-ocean thermal vents (e.g., *Taq* polymerase) enabled automation of PCR. These polymerases can withstand hours at temperatures near the boiling point of water, enabling the researcher to add them just once at the beginning of the PCR. A programmable heating unit manages the temperature shifts for a specified number of cycles. At the end of these cycles, the DNA fragment defined by the primers is visualized by agarose gel electrophoresis. It has been amplified a thousand- or million-fold, all without passage through cells!

PCR has many uses. PCR products can be cloned or sequenced. Specific known genetic mutations can be detected by PCR, even prenatally. PCR is also used to analyze forensic evidence from crime scenes. Various infectious agents, such as human immunodeficiency virus (HIV), can be identified in tissue samples through PCR. In contrast to RFLP analysis, which requires relatively large amounts of DNA, PCR is exquisitely sensitive: It can amplify DNA from a single cell of an embryo, from tiny samples of crime evidence, and also from sources of poor-quality (degraded) DNA such as Egyptian mummies, woolly mammoths, and amber-preserved insects. PCR analysis of the latter enables comparisons of the relatedness between ancient and modern organisms. Sensitivity has drawbacks, however; it is easy to contaminate PCRs with DNA that is not desired as a template.

A sequence tagged site (STS) is a site on a chromosome that yields a unique PCR fragment, usually 200 to 500 bp long. The length of the PCR fragment varies among individuals. STSs serve as physical markers in exactly the same way as RFLPs. The advantages of STS over RFLP analysis are the speed and sensitivity of the PCR-based procedure. STS analysis is yet another way to detect linkage between genes.

Topic Test 5: Polymerase Chain Reaction and Sequence Tagged Sites

True/False

1. PCR can be used to amplify uncloned sequences if some flanking (or nearby) sequence is known in order to manufacture the primers.
2. PCR uses a special thermostable DNA polymerase.

Multiple Choice

3. Which statement concerning STSs is not true?
   a. They can be linked to genes.
   b. STS detection is faster than RFLP analysis.
   c. STS detection requires less DNA than RFLP analysis.

d. They are on physical maps and genetic maps.
e. They give visible phenotypes.

4. Which one of the following is not a legitimate use of PCR?
   a. Amplify a completely uncharacterized gene from genomic DNA
   b. Clone the PCR products
   c. Sequence the PCR products
   d. Screen for genetic or infectious diseases
   e. Identify related individuals
5. What does *not* contribute to PCR efficiency?
   a. Automation
   b. Excess primers
   c. Excess template
   d. Sensitivity
   e. Thermostable polymerase

Short Answer

6. Why does PCR amplification produce DNA fragments of one size?
7. With appropriate primers, PCR-amplified products differ between individuals, as do RFLPs. An accident in a maternity ward resulted in a mix-up of the identities of three babies. Explain how PCR might be used to identify the correct infant for each set of parents.

Topic Test 5: Answers

1. **True.** Primers are necessary to perform PCR, so some sequence must be known on each side of the region whose amplification is desired.
2. **True.** The denaturation of the template DNA occurs at high temperature, making a heat-resistant DNA polymerase necessary for PCR. *Taq* polymerase is one such DNA polymerase that retains activity even after extended periods of time at very high temperature.
3. **e.** STSs can only be detected by DNA analysis.
4. **a.** If the gene is unknown and not cloned, there will be no primers known to amplify it specifically.
5. **c.** PCR is extremely sensitive. Excess template is not an important feature for efficiency. Efficiency entails using the smallest amount of DNA required to get a result.
6. Most PCR products do not contain sequence beyond (5′ of) the primers. Consequently, when these products are used as templates in later cycles, the new strand must end at that primer. Thus, together the primers define the ends of the PCR products.
7. But this isn't. STSs can be used to type all babies and parents. The babies' STS alleles must have been inherited from their parents, according to Mendel's principle of segregation.

IN THE CLINIC

Many parts of an organism's genome are the same, or nearly so, in all individuals of a species. Some parts, however, change frequently over the course of many generations so that populations contain many differing versions of those genome regions. **Variable number tandem repeats (VNTRs)** are an example of this kind of sequence in humans. The name comes from the arrangement of these sequences. Each VNTR consists of a short nucleotide sequence (approximately 15–100 bp) repeated a variable number of times, with the repeats strung together like boxcars on a train. Digestion of genomic DNA with an enzyme that does not cleave within the repeats liberates each locus of repeats as a single fragment. These can be detected by Southern blotting (see Topic 3).

Alternatively, PCR can be used to generate the fragments from small samples of DNA, such as those found at crime scenes. VNTR-containing fragments are typically between 1 and 5 kb, and there are one or many of them (i.e., many bands) per blot, depending on the type of probe used. Furthermore, the number of repeats varies between individuals and therefore the size of the fragments generated by digestion or PCR varies, as well. Because of this, VNTRs are highly polymorphic in the population, meaning there are many different forms (or alleles). The blot resembles a bar code (UPC symbol) on a grocery item.

VNTRs are heritable. Related individuals have similar patterns and unrelated people are unlikely to have similar VNTRs. This property is VNTR's claim to fame: It is used to identify lost relatives and suspects in crimes. The inheritance of VNTR alleles has been exploited to identify fathers in cases of uncertain paternity. It has also been used to reunite children separated from their parents by immigration, adoption, kidnapping, war, and other causes. The remains of the Russian royal family were identified by DNA analysis and comparison to that of living relatives. However, the most famous use of VNTRs may be in forensic applications. A match between a suspect's VNTR pattern ("DNA profile") and that found in blood or semen left at a crime scene strongly suggests guilt. Likewise, failure of suspects' DNA profiles to match that of semen evidence in rape cases can be used to show innocence. These conclusions are possible because of the exceedingly remote possibility that any two randomly selected persons will have the same VNTR profile when multiple loci are analyzed.

DEMONSTRATION PROBLEM

Question: The family shown in the pedigree has an autosomal dominant disease (affected individuals are represented by filled symbols). The gene for this disease may be on chromosome 2. All members of this family were tested for three RFLPs that map to this chromosome; their relative positions are shown. The results are shown below the pedigree and each lane corresponds to the person directly above it. Although the number of individuals is insufficient to derive mapping data, does one of the RFLPs appear to be closer than the others to the disease gene? Explain your reasoning. Assign allele designations, and indicate the most likely chromosomal configurations of both parents for RFLP B.

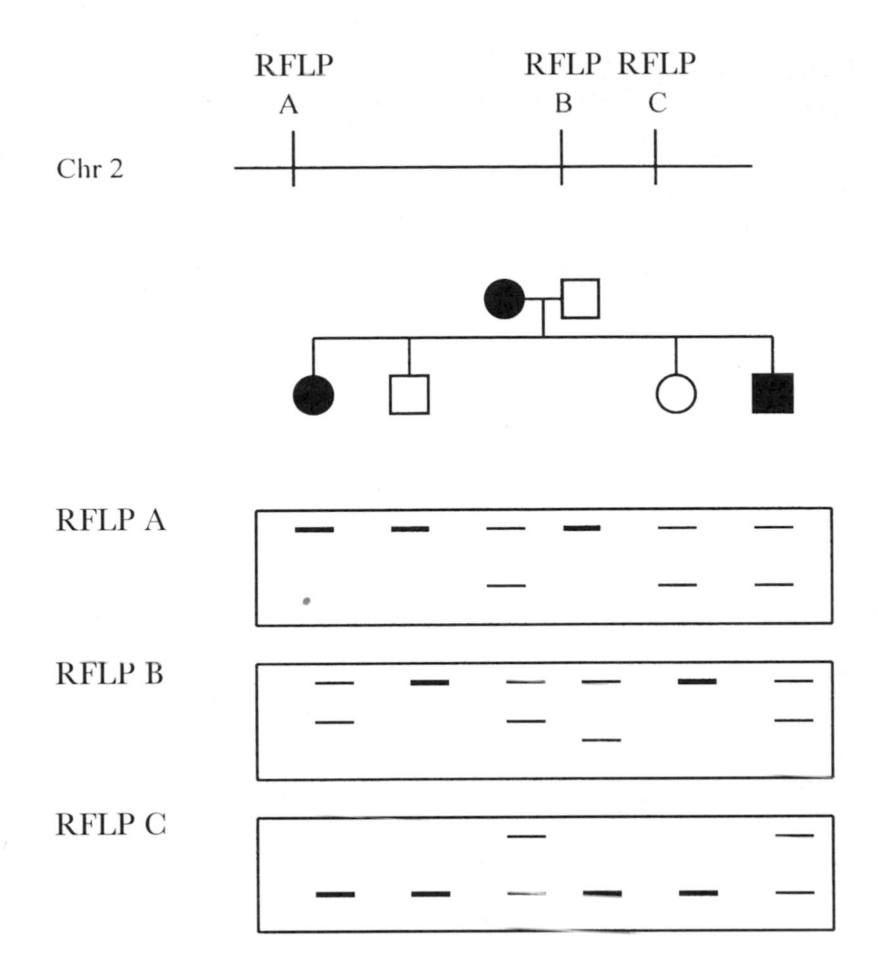

Answer: RFLP B appears to be closest to the disease gene. RFLP A shows independent assortment. RFLP C shows a crude estimate of 25% recombination: One-half of affected offspring have the mother's higher band and both unaffected offspring have the mother's lower band (the mother's higher band appears to be *cis* to the dominant disease allele). Thus, 3/4 appear to be parental and 1/4 appears to be recombinant.

The mother has two alleles of RFLP B, represented as higher and lower bands. RFLP B shows no recombination, presumably due to the tiny sample size (2/2 affected offspring have the mother's lower band and 2/2 unaffected offspring have the mother's higher band). The disease allele is *cis* to the lower RFLP allele.

Chapter Test

True/False

1. The Human Genome Project is determining the complete sequence of humans' DNA.
2. Libraries are collections of assorted sequences from a particular organism's DNA.
3. RFLPs are physical entities that cannot be mapped in genetic crosses.
4. Vectors are small, replicating DNAs used for manipulating fragments of foreign DNA.
5. PCR-based analysis of STS DNA is faster and more sensitive than RFLP analysis.

6. RNase H synthesizes cDNA.
7. PCR requires high-quality DNA templates.
8. Cohesive ends are single-stranded overhangs generated by a restriction endonuclease's asymmetric cleavage.
9. Southern blots are used to identify RFLPs.

Multiple Choice

10. What enzyme joins the ends of digested DNA?
    a. Reverse transcriptase
    b. DNA polymerase
    c. Restriction endonuclease
    d. DNA ligase
    e. RNase H
11. Which of the following is not an established technique for sequencing DNA?
    a. Products are terminated by fluorescent ddNTPs
    b. Chemical cleavage of an end-labeled molecule
    c. Products are terminated by ddNTPs and labeled by radioactive ddNTPs
    d. Products are terminated by ddNTPs and labeled by radioactive primer
    e. Products are chemically cleaved and labeled by radioactive primer
12. Practically speaking, specific genes can be identified in libraries by all of the following except
    a. complementation of a null mutation.
    b. characteristic restriction fragments.
    c. the gene product's activity.
    d. similarity to a previously identified fragment of the gene.
    e. sequencing of individual clones.
13. A fragment of yeast DNA is inserted into the ampicillin-resistance gene of a vector that also has a gene for tetracycline resistance. What is the phenotype of a bacterial cell that contains this cloned yeast DNA?
    a. Ampicillin sensitive, tetracycline sensitive
    b. Ampicillin sensitive, tetracycline resistant
    c. Ampicillin resistant, tetracycline sensitive
    d. Ampicillin resistant, tetracycline resistant
    e. None of the above
14. Reverse transcriptase
    a. is an RNA-dependent DNA polymerase.
    b. degrades RNA in RNA-DNA hybrids.
    c. is a DNA-dependent RNA polymerase.
    d. synthesizes both strands of cDNA.
    e. creates genomic libraries.
15. A woman is heterozygous for the recessive allele causing color-blindness (*cb*) and for a RFLP 5 m.u. away. Her allele configuration is *cb A/CB a*, where *A* and *a* are alleles of

the RFLP, and *CB* is the wild-type allele of the color-blindness gene. What percentage of her gametes will contain *cb* and the *a* RFLP allele?

a. 2.5%
b. 5%
c. 25%
d. 45%
e. 47.5%

Short Answer

16. What enzyme activities and DNAs are required to clone a fragment of mouse DNA?

17. A linear DNA fragment is restriction mapped with the enzymes *Eco*RI and *Bam*HI. Digestion with *Eco*RI yields fragments of 3 kb, 4 kb, and 5 kb. Digestion with *Bam*HI yields fragments of 2.5 kb, 3.5 kb, and 6 kb. Simultaneous digestion with both enzymes yields fragments of 1 kb, 1.5 kb, 2 kb, 2.5 kb, and 5 kb. Draw a restriction map consistent with these data.

18. Mary is heterozygous for a mutation that causes a dominant phenotype; her husband is normal. She wants to know if the fetus she carries is also affected. The mutation itself (not a linked RFLP) can be directly detected by PCR, where the wild-type and mutant alleles give products of differing size. This analysis generated the picture below. Does the fetus have the mutant allele? Explain.

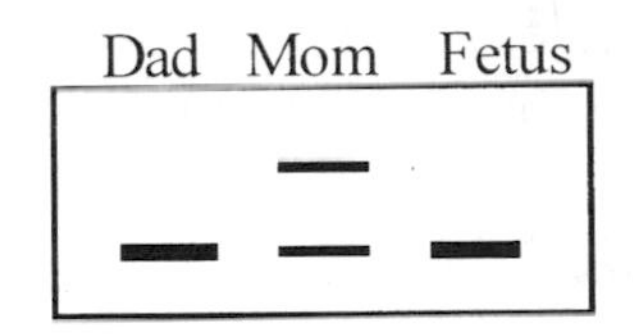

Essay

19. Describe the way mRNA is made into a cDNA library.

20. How are dideoxyribonucleotides used in sequencing?

Chapter Test Answers

1. **T** 2. **T** 3. **F** 4. **T** 5. **T** 6. **F** 7. **F** 8. **T** 9. **T** 10. **d** 11. **e** 12. **e**

13. **b** 14. **a** 15. **a**

16. In addition to mouse DNA, a vector such as a cosmid is needed. A restriction endonuclease will be needed to generate the same cohesive ends in the vector and in the mouse DNA. DNA ligase is needed to covalently join the vector and mouse DNAs.

17. In the figure, B represents *Bam*HI cleavage sites; E represents *Eco*RI cleavage sites. Numbers are sizes of fragments in kilobases.

B E B E
2.5 1.5 2 1 5 kb

18. The fetus does not have the mutant allele. Its father is normal so the single band PCR that generates from his DNA is from the wild-type allele. Mary is heterozygous, as can be seen in the PCR products from her DNA. The lower band must be from the wild-type allele, identified as such by her husband's PCR product. The fetus has a PCR product from only the wild-type allele, so it must be homozygous wild type.
19. The mRNA is isolated from a cell type in which the gene of interest is expressed, or the researcher is interested in the total mRNA in the cell type. Reverse transcriptase is used to synthesize a DNA copy of the mRNA templates. RNase H degrades the RNA partner in the RNA-DNA hybrid while DNA polymerase I uses the partially degraded RNAs as primers for synthesis of the second DNA strand. The resulting double-stranded, blunt-ended cDNAs are cloned into appropriate vector(s). This diverse collection of cDNA-containing vectors is maintained as a mixture: a **cDNA library**.
20. Dideoxyribonucleotides (ddNTPs) lack the 3′ hydroxyl moiety on ribose to which the next (3′) nucleotide would be added. Since no other nucleotide can be attached to the 3′ end, incorporation of ddNTPs causes chain termination. Sequencing reactions typically contain one kind of ddNTP in small quantities and the four regular nucleotides that are used to synthesize a new strand of DNA from the template whose sequence is desired. The ddNTPs are randomly and infrequently incorporated, generating a population of differently sized molecules, all having the same last (3′) nucleotide. By determination of the lengths of the terminated fragments (by gel electrophoresis), the positions of the ddNTP (and its cognate nucleotide) are discovered. Four sequencing reactions are performed per template so that all four ddNTPs are used to determine the positions of all four normal nucleotides in the sequence. In fluorescent/PCR sequencing, the four reactions occur in one tube and are analyzed on one lane of a gel.

Check Your Performance:

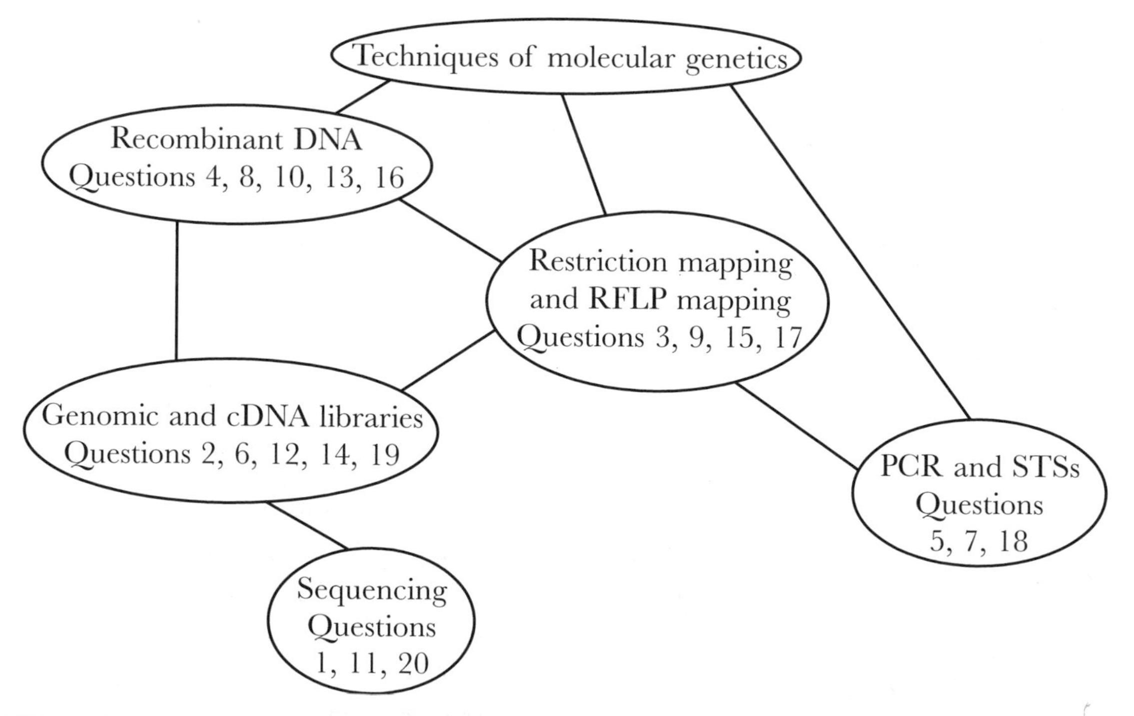

Use this chart to identify weak areas, based on the questions you answered incorrectly in the Chapter Test.

Population Genetics

We previously considered the results of specific single matings in formulating genetic theories. In fact, the evidence for Mendel's principles are the results of such matings, repeated many times to generate statistically significant numbers. This chapter seeks to extend Mendel's principles to whole groups of matings, between individuals of varying genotypes. The natural outgrowth of such a change in scale is to contemplate the change in genetic structure of populations over successive generations, that is, the origins of species.

ESSENTIAL BACKGROUND

- **Mendel's principles and genotypic frequencies (Chapter 1)**
- **Frequencies, chi-square analysis, and random fluctuations (Chapter 2)**
- **Incomplete penetrance (Chapter 3)**
- **Sex linkage (Chapter 3)**
- **Mutation and reversion (Chapter 6)**

TOPIC 1: GENOTYPIC AND ALLELIC FREQUENCIES

KEY POINTS

✓ *What are populations? genetic structure? gene pools? allelic frequencies?*

✓ *How are allelic frequencies determined?*

A **population** is a group of individuals of a particular species. The commonly accepted definition of what constitutes a population is all individuals of a species that are capable of random mating (vis-à-vis timing, geography, etc.). The **genetic structure** of a population is all of the alleles present in the population, their frequencies, and how these alleles are distributed among genotypes of individuals. The complete collection of a population's alleles is also called its **gene pool**. There is nothing about a population's genetic structure that must remain constant; the processes that affect genetic structure are the focus of later topics. The genetic structure can vary over time or over geographic space. For example, one population of crows may have a different frequency of heterozygotes than another population separated from it by a lack of interbreeding. Frequencies of heterozygotes and homozygotes are **genotypic frequencies** and are used extensively in population genetics. The genotypic frequency for dominant homozygotes is the number of dominant homozygotes, divided by the total number of individuals in that population. Similar calculations give the frequencies of recessive homozygotes and heterozygotes. For example, if a population consists of 530 *AA*, 90 *Aa*, and 380 *aa* individuals, the genotypic frequencies are as follows: f(*AA*) = 0.53, f(*Aa*) = 0.09, and f(*aa*) = 0.38. The frequencies of all of the homozygotes and heterozygotes in a particular population will sum to 1.

Allelic frequencies are the sum of the numbers of each allele present in homozygotes and heterozygotes, divided by the total number of alleles in the population. These frequencies also must sum to 1. Keep in mind that for most species, there are two alleles of each autosomal gene per individual, so the total number of alleles will be twice the number of individuals. By convention, the frequency of the dominant allele is given the symbol p and the frequency of the recessive allele is symbolized by q. If there are only two alleles of a particular gene, then $p + q = 1$, meaning the sum of the frequencies of all the dominant and recessive alleles in the population equals all the alleles in the population. The allelic frequencies for the previous example of 530 *AA*, 90 *Aa*, and 380 *aa* individuals are calculated as follows. The number of *A* alleles in homozygotes is 2×530 (because each homozygote has two alleles), or 1,060, and the number present in heterozygotes is 90 (one for each of the 90 *Aa* individuals). The sum of all *A* alleles (1,150) is divided by the total number of alleles ($2 \times 1{,}000$) to obtain $p = 0.575$. A similar calculation gives the frequency of the *a* allele, $q = 0.425$. Alternatively, these frequencies could have been determined from the genotypic frequencies: $p = f(AA) + 1/2\ f(Aa) = 0.53 + 1/2\ (0.09) = 0.575$ and $q = f(aa) + 1/2\ f(Aa) = 0.38 + 1/2\ (0.09)$. Notice that the values of the two allelic frequencies are mutually dependent; since $p + q = 1$; once p is known, q can be calculated from $q = 1 - p$. If the population has more than two alleles for a gene, additional terms are added to this equation; for example, for three alleles, $p + q + r = 1$.

Allelic frequencies of X-linked genes are a bit more complicated. Females have two alleles of these genes, but males have only one, so special steps are taken to account for this in the calculation. For example, red-green color blindness is caused by an X-linked recessive allele, X^c. Let X^+ be the wild-type allele. In a particular population, there are 810 X^+X^+, 180 X^+X^c, 10 X^cX^c, 900 X^+Y, and 100 X^cY individuals. The p value is two times the number of X^+ alleles in homozygous females, plus the number in heterozygous females, plus the number in hemizygous males, divided by the total number of alleles: $[(2 \times 810) + 180 + 900]/3{,}000 = 0.90$. Therefore, $q = 1 - 0.90 = 0.10$.

For practical reasons, frequencies are usually determined from sampling a population, rather than counting the entire population. Together allelic frequencies and genotypic frequencies quantitatively describe a population's genetic complexity. To accurately represent a population, both sets of frequencies are necessary because for one set of allelic frequencies, there are many possible sets of genotypic frequencies. For example, the allele frequencies $p = f(A) = 0.6$ and $q = f(a) = 0.4$ are true for both of the following populations:

population I f(*AA*) = 0.36; f(*Aa*) = 0.48; f(*aa*) = 0.16

population II f(*AA*) = 0.60; f(*Aa*) = 0.00; f(*aa*) = 0.40

Furthermore, the set of frequencies is accurate for the generation of the population for which they were determined but *may not* be true of the next generation or the next population. The only way to know whether the frequencies are unchanged is to sample the other or later population.

Topic Test 1: Genotype and Allelic Frequencies

True/False

1. Allelic frequency, once determined, will be true of all subsequent generations of the same population.

2. A gene pool is all of the genotypes present in the population.
3. Genetic complexity of a population is described by genotypic and allelic frequencies.

Multiple Choice

4. What are the genotypic frequencies in a population consisting of 312 *AA*, 361 *Aa*, and 242 *aa* individuals?
   a. f(*AA*) = 0.682; f(*Aa*) = 0.790; f(*aa*) = 0.528
   b. f(*AA*) = 0.312; f(*Aa*) = 0.446; f(*aa*) = 0.242
   c. f(*AA*) = 0.3; f(*Aa*) = 0.4; f(*aa*) = 0.3
   d. f(*AA*) = 0.312; f(*Aa*) = 0.361; f(*aa*) = 0.242
   e. f(*AA*) = 0.341; f(*Aa*) = 0.395; f(*aa*) = 0.264
5. Populations have all but one of the following:
   a. A gene pool
   b. Genetic structure
   c. Allelic frequencies
   d. Genotypic frequencies
   e. No interbreeding within the group

Short Answer

6. In one population of marmosets, the genotypic frequencies for the *M* gene are 0.42*MM*, 0.08*Mm*, and 0.50*mm*. What are the frequencies of these alleles?
7. An X-linked recessive allele causes the blood clotting disorder hemophilia. In a particular population, there are 900 homozygous normal females, 99 heterozygous normal females, 1 hemophiliac female, 968 normal males, and 32 hemophiliac males. What are the allelic frequencies in this population?

Topic Test 1: Answers

1. **False.** Allelic frequencies can be constant, or change every few generations, or change every generation.
2. **False.** A gene pool is all of the alleles present in a population.
3. **True.** However, one set of allelic frequencies can have many possible genotypic frequencies. To be unambiguous, if allelic frequencies are given, genotypic frequencies must also be reported.
4. **e.** Total population is 915. f(*AA*) = 312/915 = 0.341; f(*Aa*) = 361/915 = 0.395; f(*aa*) = 242/915 = 0.264. The frequencies can be determined more accurately than those in answer c, which have too few significant digits.
5. **e.** Members of a population must be able to breed freely within the group.
6. $p(M) = (0.08/2) + 0.42 = 0.46$ and $q(m) = (0.08/2) + 0.50 = 0.54$.
7. There are 2,000 people and 3,000 alleles in this population. Two thousand alleles are from the 1,000 females and 1,000 are from the 1,000 males because males are hemizygous. The

frequency of the normal allele is $p = [(900 \times 2) + 99 + 968]/3{,}000 = 0.96$ and $q = [(1 \times 2) + 99 + 32]/3{,}000 = 0.04$.

TOPIC 2: HARDY-WEINBERG LAW

KEY POINTS

✓ *What is Hardy-Weinberg equilibrium?*

✓ *What are the conditions for Hardy-Weinberg equilibrium?*

✓ *What is a population Punnett square?*

In 1908, Godfrey Hardy and Wilhelm Weinberg independently proposed a mathematical model for how Mendel's principles of inheritance apply to populations to produce allelic frequencies and genotypic frequencies. The model, now known as the **Hardy-Weinberg law**, states that given a set of five assumptions, allelic frequencies do not change over time and $f(AA) = p^2$, $f(Aa) = 2pq$, and $f(aa) = q^2$. The assumptions are that (1) the population is large (so that genetic drift is less), (2) the population randomly mates, and that it is not experiencing (3) mutation, (4) migration, or (5) natural selection (see below).

For a particular pair of allelic frequencies (p and q), there is one set of genotypic frequencies that constitutes **Hardy-Weinberg equilibrium**. Once a population is in Hardy-Weinberg equilibrium, all subsequent generations will remain in equilibrium as long as the five assumptions continue to be true. This equilibrium reflects unchanging allelic frequencies, which are a direct result of meiosis and random mating. Meiosis produces gametes that proportionally represent the alleles present in the parent. Random mating produces genotypes at frequencies given by multiplying the allelic frequencies of the contributing gametes. This concept is implicit in the **population Punnett square**. Consider the following example:

			All females' gametes			
			A	$p = 0.7$	*a*	$q = 0.3$
All males' gametes	*A*	$p = 0.7$	*AA*	0.49	*Aa*	0.21
	a	$q = 0.3$	*Aa*	0.21	*aa*	0.09

This Punnett square represents all possible matings in a population whose allelic frequencies are $p = 0.7$ and $q = 0.3$. According to the square, 0.7 of the gametes in the population are *A* for both males and females and 0.3 of the gametes are *a* for both males and females. The boxes within the square contain all of the progeny expected in the next generation under Hardy-Weinberg equilibrium. Notice that the frequency of each type of progeny is a product of the frequencies of the alleles that form that genotype. Furthermore, these products can be generalized as $f(AA) = p^2$, $f(Aa) = 2pq$, and $f(aa) = q^2$, and the sum of these is 1. In this specific example, the parental generation is in Hardy-Weinberg equilibrium if their genotypic frequencies are 0.49 *AA*, 0.42 *Aa*, and 0.09 *aa*. If they are not, as the result of being in violation of one or more of the assumptions, this Punnett square demonstrates the time it takes for the population to reach equilibrium once the assumptions have been met: one generation—the one that is the children of the generation in which the assumptions have been met again.

Violation of any of the assumptions causes disruption of allelic frequencies or disruption of genotypic frequencies or both. Hence, if a population is determined not to be in Hardy-Weinberg equilibrium, the cause can be investigated. What might be found is the reason for evolutionary change. Migration and mutation can disrupt allelic frequencies by introducing more alleles of one or another type (see Topic 4). Natural selection removes alleles nonrandomly from the population (see Topic 7). The frequencies of randomly mated pairs are determined by multiplying the frequencies of the two genotypes involved. If the actual frequencies of matings do not match this prediction, the population is said to be nonrandomly mating (see Topic 5). Finally, genetic drift is the random fluctuation of allelic frequencies that affects *small* populations to a much greater extent than *large* populations (see Topic 6). Whenever any of these processes occurs to a population, allelic frequencies can change and Hardy-Weinberg equilibrium will be lost. The generation born after the disruption has ceased will again be in Hardy-Weinberg equilibrium. However, the allelic and therefore genotypic frequencies in this population may be different from those of the earlier generation in equilibrium.

There are notable practical ramifications of the Hardy-Weinberg law that may surprise you. First, if a population is in Hardy-Weinberg equilibrium, dominant alleles will not "naturally" increase in frequency. Recessive alleles will not "naturally" die out. Either of these events could happen as a result of natural selection or genetic drift, but this would cause the population to fall out of Hardy-Weinberg equilibrium. Secondly, the frequency of heterozygotes in the population is largest when $p = 0.5$ (and $q = 0.5$). Thirdly, when p or q is small, most of these rare alleles will be present in heterozygotes, rather than in homozygotes. This last point should be familiar from pedigree analysis (see Chapter 3).

Topic Test 2: Hardy-Weinberg Law

True/False

1. The frequency of the homozygous dominant genotype is always given by p^2.
2. If q is small, the recessive alleles will mostly exist in heterozygotes.

Multiple Choice

3. All of the following can disrupt Hardy-Weinberg equilibrium except
   a. migration.
   b. random mating.
   c. natural selection.
   d. genetic drift.
   e. mutation.
4. If the conditions are met, how long does it take to establish Hardy-Weinberg equilibrium?
   a. It depends on which condition was not met before.
   b. It takes until the generation in which conditions are met.
   c. It takes until the offspring of the generation in which conditions are met.
   d. It takes until the second generation after conditions are met.
   e. It takes an indeterminate time.

Short Answer

5. How do Mendel's principles relate to the Hardy-Weinberg law?

6. Draw a population Punnett square to calculate the genotypic frequency of a population that is in Hardy-Weinberg equilibrium and has allelic frequencies $p = f(A) = 0.4$ and $q = f(a) = 0.6$.

Topic Test 2: Answers

1. **False.** This statement is true if and only if the population is in Hardy-Weinberg equilibrium.
2. **True.** To see why this is true, pick a small number for q (e.g., $q = 0.001$). Determine p, and use these to calculate $2pq$ and q^2.
3. **b.** Random mating is necessary for Hardy-Weinberg equilibrium. Nonrandom mating changes genotypic frequencies.
4. **c.** Hardy-Weinberg equilibrium will be observed in the first generation to be created in the absence of all conditions that prevent Hardy-Weinberg equilibrium.
5. Mendel's principles arise from the mechanics of meiosis. According to Mendel's principles, the frequencies of gametes reflect their frequencies in the individual. By extension, the frequencies of alleles in the gametes of a population reflect their frequencies in the previous generation under Hardy-Weinberg equilibrium.
6.

		A	$p = 0.4$	a	$q = 0.6$
A	$p = 0.4$	$f(AA)$	0.16	$f(Aa)$	0.24
a	$q = 0.6$	$f(Aa)$	0.24	$f(aa)$	0.36

So $f(AA) = 0.16$, $f(Aa) = 0.48$, and $f(aa) = 0.36$.

TOPIC 3: TESTING FOR HARDY-WEINBERG EQUILIBRIUM

KEY POINTS

✓ *How do we recognize whether a population is in Hardy-Weinberg equilibrium?*

✓ *How do we determine genotypic frequencies when heterozygotes are indistinguishable from dominant homozygotes?*

✓ *What special considerations must be made for X-linked loci? for multiple alleles?*

A significant use of the Hardy-Weinberg law is to determine whether a population is in equilibrium. If it is not, the observed deviations may suggest which assumption(s) of the law is (are) being violated. Populations that are in equilibrium will have the expected genotypic frequencies (p^2, $2pq$, and q^2) for the observed allelic frequencies. To test whether a population is in Hardy-Weinberg equilibrium, we first determine the allelic frequencies p and q from the population's genotypes. Next, we use these allelic frequencies to calculate *expected* genotypic frequencies at Hardy-Weinberg equilibrium, using the formulae $f(AA) = p^2$, $f(Aa) = 2pq$, and $f(aa) = q^2$. Finally, we convert the genotypic frequencies to predicted numbers for each genotype, scaled to the size of the observed population.

The critical test of Hardy-Weinberg equilibrium is to compare by chi-square analysis the actual population distribution with the predicted distribution that assumed Hardy-Weinberg equilibrium. This process is demonstrated for the following population: 189 *AA*, 202 *Aa*, and 609 *aa* individuals. Is this population in Hardy-Weinberg equilibrium? For this population, $p = [(2 \times 189) + 202]/2{,}000 = 0.29$ and $q = [(2 \times 609) + 202]/2{,}000 = 0.71$, so the equilibrium genotypic frequencies are $f(AA) = p^2 = (0.29)^2 = 0.084$; $f(Aa) = 2pq = 2 \times 0.71 \times 0.29 = 0.412$; and $f(aa) = q^2 = (0.71)^2 = 0.504$. The sum of 189 + 202 + 609 = 1,000 individuals in the original population. Therefore, expected numbers are 1,000 × 0.084 = 84 *AA*, 1,000 × 0.412 = 412 *Aa*, and 1,000 × 0.504 = 504 *aa*. The observed numbers are 189 *AA*, 202 *Aa*, and 609 *aa*, which obviously differ significantly from the Hardy-Weinberg equilibrium prediction ($\chi^2 = 260$, df = 1 because n = 2 since the third genotypic class is not independent; once the other two are known, there is only one possible value for the third, so $p <<< 0.05$). Thus, this population is not in Hardy-Weinberg equilibrium. Note that the allelic frequencies were determined by counting alleles; this is the preferred method of testing for Hardy-Weinberg equilibrium.

Often, however, it is impossible or impractical to distinguish heterozygotes from dominant homozygotes. A different approach is needed in these cases. First, *assume* the population is in Hardy-Weinberg equilibrium so that q can be determined by taking the square root of the recessive homozygote frequency. Then proceed to determine frequencies for the other two genotypes. We cannot test whether this population is in Hardy-Weinberg equilibrium because we already assumed it is, in order to calculate frequencies. Yet we make this assumption in order to *estimate* the frequency of heterozygotes when we have no other way of identifying them.

If there are more than two alleles of a gene, for example, the ABO blood group in humans, the genotypic frequencies at Hardy-Weinberg equilibrium are given by $p^2 + 2pq + q^2 + 2pr + 2qr + r^2 = 1$. Care must be taken in this situation to avoid confusing the plethora of genotypes. Where A^1, A^2, and A^3 are three alleles of a single gene, p^2 is A^1A^1, $2pq$ is A^1A^2, q^2 is A^2A^2, $2pr$ is A^1A^3, $2qr$ is A^2A^3, and r^2 is A^3A^3. Another tricky situation is that of X-linked alleles. The expected genotypic frequencies for females are calculated as for autosomal loci, but the frequencies for males are the same as the allelic frequencies, p and q. For example, the X-linked recessive hemophilia allele is rare: $q = 0.0001$. Consequently, hemophiliac males are expected at the Hardy-Weinberg equilibrium frequency $q = 0.0001$ and hemophiliac females at the frequency $q^2 = 0.00000001$.

Topic Test 3: Testing for Hardy-Weinberg Equilibrium

True/False

1. A population whose genotypic frequencies do not match the expectation based on their allelic frequencies and an assumption of Hardy-Weinberg equilibrium are not in Hardy-Weinberg equilibrium.
2. For X-linked genes, genotypic frequency for males is either p or q.
3. Hardy-Weinberg equilibrium testing is only for two-allele genes.

Multiple Choice

4. The MN blood group is specified by two codominant alleles of a single gene. Homozygotes are blood type M or N and heterozygotes are blood type MN. There is an unequal geographic distribution of these alleles such that a group of Alaskan Eskimos contains 8,281 M-type, 1,638 MN-type, and 81 N-type individuals. An equal-size group of Australian aborigines consists of 324 M-type, 2,952 MN-type, and 6,724 N-type individuals. Which population is in Hardy-Weinberg equilibrium?

a. The Alaskan Eskimos
b. The Australian aborigines
c. Neither
d. Both
e. Impossible to tell

5. Which one of the following steps is not part of the procedure for determining whether a population is in Hardy-Weinberg equilibrium?
   a. Calculate equilibrium frequencies of genotypes from allelic frequencies.
   b. Compare expected numbers of each genotype at Hardy-Weinberg equilibrium to the observed population.
   c. Convert genotypic frequencies to predicted number of each genotype.
   d. Obtain p and q by counting alleles in the population.
   e. Obtain q by taking the square root of the frequency of the recessive homozygotes.

Short Answer

6. Red and white snapdragons are homozygous for incompletely dominant alleles of a single gene. The heterozygote is pink. A field of snapdragons contains 740 red, 240 pink, and 20 white flowering plants. Is this population in Hardy-Weinberg equilibrium?
7. Color blindness is an X-linked recessive disorder. In a particular population of 5,000 individuals, there are 75 affected males and 2,425 normal males. How many affected, carrier, and homozygous normal females are present if the population is in Hardy-Weinberg equilibrium?

Topic Test 3: Answers

1. **True.** If the observed and expected genotypic frequencies do not match, the population is not in equilibrium. The comparison is normally performed with numbers rather than frequencies to avoid ignoring (or being misled by) random fluctuations related to sample size and to enable significance testing by chi-square analysis.
2. **True.** Males have a single X chromosome so they have just one allele of X-linked genes.
3. **False.** The formula $p^2 + 2pq + q^2 + 2pr + 2qr + r^2 = 1$ accounts for all possible genotypes of a three-allele gene.
4. **d.** Genotypic frequencies can be very different from population to population and yet each population can still be in equilibrium! Also notice that although both are in Hardy-Weinberg equilibrium, the equilibria have been reached at different frequencies for *the same alleles*. In fact, alternate alleles predominate in the two populations. For the Alaskan Eskimos, $p(L^M) = 0.91$ and $q(L^N) = 0.09$. For the Australian aborigines, $p(L^M) = 0.18$ and $q(L^N) = 0.82$.
5. **e.** This action *assumes* the population is in Hardy-Weinberg equilibrium, so using it to test for equilibrium would invoke circular reasoning.
6. Allelic frequencies are calculated by counting the alleles in the population. For the red allele, $[(2 \times 740) + 240]/2{,}000 = 0.86$. For the white allele, $[(2 \times 20) + 240]/2{,}000 = 0.14$.

Expected genotypic frequencies are 0.74 red, 0.24 pink, and 0.02 white. Expected numbers of plants are 740 red, 240 pink, and 20 white. The population appears to be in Hardy-Weinberg equilibrium.

7. This trait is X-linked so $q = 75/2{,}500$, or 0.03. $p = 1 - q = 1 - 0.03 = 0.97$. The frequency for normal females is $p^2 = 0.94$, or there are approximately 2,350. For carrier females the frequency is $2pq = 0.058$, or the number is approximately 145. For affected females the frequency is $q^2 = 0.0009$, or the number is approximately 2.

TOPIC 4: MUTATION AND MIGRATION

KEY POINT

✓ *What effects do mutation and migration have on a population's genetic structure?*

Mutation is a continuing source of new alleles in a population. As such, it makes evolution possible. If a population is all *AA*, occasionally *a* alleles will be created, and if the population has abundant *A* and *a* alleles already, mutation will interconvert some *A* to *a* and some *a* to *A*. Mutation does not happen because one allele becomes scarce; it happens at a relatively constant low rate. Therein lies the limitation of mutation as an evolutionary force; it would have significant effects on allelic frequencies if it were not so rare. The typical mutation frequency for a human gene (based on autosomal dominant and X-linked recessive mutations) is approximately 1 mutation per 10^5 to 10^6 meioses. Consider the mutation of *A* to *a* where initially there are no *a* alleles in the population ($q = 0$ and $p = 1$). The frequency of the alleles after one generation of mutation is given by the equations: $p_1 = p_0 - \mu p_0$ and $q_1 = 1 - p_1$, where p_0 and p_1 are the old and new frequencies of the *A* allele, respectively; μ is the mutation rate from *A* to *a*; and q_1 is the new frequency of the *a* allele. If the mutation rate is 1×10^{-5}, the new frequency of the *A* allele is $p_1 = 1 - 1 \times (1 \times 10^{-5}) = 0.99999$. The new frequency of the *a* allele is $q_1 = 1 - p_1 = 1 - 0.99999 = 1 \times 10^{-5}$. Thus, allelic frequencies are only very slightly changed, yet the variability within the population has increased. Furthermore, the *a* alleles are subject to reversion back to *A* alleles in subsequent generations. Eventually an equilibrium will be reached between forward and back mutation, but it will not be Hardy-Weinberg equilibrium. As an evolutionary process, mutation is too rare to account for the rapid evolution of species. However, mutation supplies the starting material that evolutionary processes work on.

Migration is the movement of a group of individuals into another population with the potential to alter the allelic frequencies of that population and perhaps to introduce new alleles into it. The magnitude of the effect of the migrating group is proportional to its size relative to the recipient population and to the degree of similarity between their allelic frequencies. If the migrating population is small compared to the recipient population, the migration will have a smaller effect on the allelic frequencies in the new population than if there are many migrants. Large migrating populations generally have the greatest effect. If the allelic frequency in the migrating population is very similar to that in the recipient population, the migration will have a smaller effect on the allelic frequencies in the new population than if the allelic frequencies are dissimilar. Very dissimilar migrants generally have the greatest effect.

These trends can be seen mathematically in the equation that models the overly simplistic situation of one group of individuals migrating into an existing population: $\Delta p = \mathrm{m}\,(p_{\mathrm{m}} - p)$, where Δp is the change in p from the resident population to the new population; m is the migration coefficient, which is the proportion of migrants in the new population; p_m is p in the migrating population; and p represents the allelic frequency in the recipient population before the migration. If a group of 50 sparrows with $p = 0.4$ joins an existing population of 250 sparrows with $p = 0.85$, the p of the new population is obtained through a two-step calculation. First, determine the change in p: $\Delta p = m(p_m - p) = (50/300)(0.4 - 0.85) = (0.167)(-0.45) = -0.075$. Second, adjust the p of the recipient population by this value, taking care to retain the sign of Δp: $p_{new} = p + \Delta p = 0.85 + -0.075 = 0.775$. Thus, the frequency of the *A* allele decreased as a result of the migration. This makes sense given the differing allelic frequencies, and at the same time, the recipient population has grown in size, making it more resistant to genetic drift (see Topic 6). The general effect of migration is to increase both the size of the recipient population and the genetic variation of populations by spreading alleles that may exist in one population but not in others. This essentially homogenizes populations, making them more similar to each other. Although Hardy-Weinberg equilibrium is disrupted in the generation in which the migration occurs, it can be restored in the next generation if the required conditions are again met.

Topic Test 4: Mutation and Migration

True/False

1. Mutation is responsible for the rapid evolution of species.
2. Migration causes large changes in allelic frequencies.
3. New alleles in populations can come from mutation.

Multiple Choice

4. Which one of the following is not a feature of mutation?
   a. It is rare.
   b. It disrupts Hardy-Weinberg equilibrium.
   c. It causes *a* to be converted to *A*.
   d. It has a large effect on allelic frequency.
   e. It causes increased variation.
5. Which one of the following is not an effect of migration?
   a. It increases differences between populations.
   b. It introduces new alleles into populations.
   c. It homogenizes populations.
   d. It increases the size of recipient populations.
6. Neurofibromatosis (NF) is an autosomal dominant disease characterized by multiple tumors in skin and nerves. The spontaneous mutation frequency for NF is about 7×10^{-5}. What is the new frequency of the *normal* allele after one generation if the population initially is $p = 0.00010$, $q(\text{normal}) = 0.99990$? (Assume the frequency of NF alleles reverting to wild type is 0.)
   a. 0.99990
   b. 0.99983
   c. 0.00007
   d. 0.00017
   e. 0.00010

Short Answer

7. A population of 150 pigeons lives around the state capitol building. The frequency of the Pdg^+ allele among these pigeons is 0.94. An invading group of 20 pigeons comes from a population where the *Pdg* allelic frequency is 0.66. After the invaders mix with the capitol population, what is the new allelic frequency of Pdg^+?

Topic Test 4: Answers

1. **False.** Mutation is too rare to be responsible for anything rapid. See Topic 7 for the strongest mechanism of evolution.
2. **False.** Migration has variable effects that depend on the sizes of the populations involved and their allelic frequencies.
3. **True.** An example is a population fixed ($p = 1$) for the *A* allele, in which mutation converts one *A* into *a*, a new allele.
4. **d.** Mutation is too rare to cause large alterations of allelic frequencies.
5. **a.** Migration decreases the differences between populations by transferring alleles that may be present in one but not the other population and by blending allelic frequencies.
6. **b.** For this problem, $q_0 = 0.99990$ and $\mu = 7 \times 10^{-5}$. The value μ represents the change from the normal allele (q) to the dominant disease allele (p). Revise the formula to reflect the dominance of the NF allele and so that the mutation rate is paired with the appropriate allele: $q_1 = q_0 - (q_0)\mu = 0.99990 - (0.99990)(7 \times 10^{-5}) = 0.99983$.
7. $m = 20/(20 + 150)$, $p_m = 0.66$, and $p = 0.94$. $\Delta p = m\,(p_m - p) = (20/170)(0.66 - 0.94) = -0.03$. $p_{new} = 0.94 + (-0.03) = 0.91$.

TOPIC 5: RANDOM MATING

KEY POINTS

✓ *What is inbreeding? outbreeding? assortative mating?*

✓ *What are the consequences of nonrandom mating?*

Hardy-Weinberg equilibrium assumes that mating is random with respect to *the locus under consideration* (not all loci). Often this is not true. These nonrandomly mating populations may not exhibit the genotypic frequencies predicted by Hardy-Weinberg equilibrium. Two kinds of nonrandom mating are recognized, distinguished by the number of loci affected: inbreeding/outbreeding and assortative mating.

Inbreeding is mating that occurs more frequently between genetically related individuals than is predicted by chance. The effect is to increase homozygosity of all loci with each successive generation of inbreeding, relative to Hardy-Weinberg equilibrium frequencies. An extreme example of inbreeding comes from self-fertilizing plants, which are common in the wild. Half of the progeny of a self-fertilizing heterozygote are homozygous (either *AA* or *aa*) and half are heterozygous. At the same time, self-fertilization of homozygous individuals produces exclusively homozygous progeny. Thus, the proportion of the total population that is homozygous increases in each generation, at the expense of the proportion of heterozygotes, because heterozygotes

contribute to the proportions of all three genotypes but homozygotes only contribute to their own genotype subpopulation. With each successive generation of self-fertilization, the frequency of heterozygotes diminishes by half, assuming equal fertility of the genotypes. Theoretically, the heterozygote is eventually eliminated from the population. Successive generations of inbreeding produce populations that are progressively more unlike the Hardy-Weinberg equilibrium prediction, even though allelic frequencies do not change if the population is large (therefore resistant to genetic drift; see Topic 6). Overall, inbreeding causes a decrease in heterozygosity and concomitant increase in homozygosity. This is the reason for the accumulation of deleterious as well as desirable traits in purebred (i.e., highly inbred) pets, crops, livestock, and so on.

The converse of inbreeding is outbreeding, which is also common in the wild. **Outbreeding** is mating that occurs *less* frequently between genetically related individuals than predicted by chance. Homozygotes are less likely to mate with individuals of the same genotype and so are more likely to produce heterozygous offspring. Thus, an outbred population experiences an overall *increase* in heterozygosity. Allelic frequencies may not change as a result of outbreeding or inbreeding, but Hardy-Weinberg equilibrium genotypic frequencies will not be observed, and the deviation from expected values will increase with increasing generations of outbreeding. Both inbreeding and outbreeding tend to affect many traits, because relatives are likely to have many alleles of many genes in common.

Assortative mating occurs when individuals chose each other on the basis of similarity, which is common in wild populations. Humans often engage in **positive assortative mating** for characteristics such as height, skin color, or weight, choosing mates similar to themselves. Positive assortative mating examples include people of similar height being more likely to mate than expected by chance. Because single traits are often the basis for this kind of mating, usually only those traits are affected, unlike inbreeding and outbreeding, which affect all traits. As long as the traits that are the basis for an assortative mating are not the ones being studied, the mating is considered to be random for the purpose of expecting Hardy-Weinberg equilibrium frequencies. Allelic frequencies are rarely altered by assortative mating but genotypic frequencies may or may not be affected, depending on whether the trait is genetic. For example, height has a large environmental basis, so two tall individuals who have mated assortatively may, in fact, have differing genotypes. **Negative assortative mating** is the mating of dissimilar individuals more often than would be predicted by chance. This, too, affects genotypic frequencies but not allelic frequencies and occurs frequently in natural populations.

Topic Test 5: Random Mating

True/False

1. Inbreeding is more matings between relatives than there would be by chance.
2. Assortative mating is rare.
3. Outbreeding causes many genes to become homozygous.

Multiple Choice

4. If a population consists of 25% *AA*, 50% *Aa*, and 25% *aa* individuals, how many generations of strict inbreeding will result in about 1% heterozygotes left in the population?

a. 1
b. 3
c. 6
d. 10
e. 50

5. Which of the following does not occur in wild populations?
   a. Inbreeding
   b. Positive assortative mating
   c. Outbreeding
   d. Negative assortative mating
   e. None of the above

Short Answer

6. In what way do inbred and outbred populations deviate from Hardy-Weinberg equilibrium predictions?

Topic Test 5: Answers

1. **True.** This is the definition of inbreeding.
2. **False.** Assortative mating is quite common and so long as the trait being studied is not the one that was the basis for the assortative matings, these matings still qualify as random.
3. **False.** Outbreeding means homozygotes are more likely to mate with the other homozygotes or with heterozygotes, both of which increase the proportion of heterozygotes in the offspring.
4. **c.** Inbreeding causes a 50% reduction in the numbers of heterozygotes per generation. The first generation after the one cited would contain 25% heterozygotes, for example. The second would contain 12.5% heterozygotes, and so on.
5. **e.** All of these mechanisms are known to occur in the wild.
6. Both kinds of nonrandom mating produce alterations in genotypic frequencies but do not change allelic frequencies. Inbred populations contain greater numbers of homozygotes than Hardy-Weinberg equilibrium predicts. Outbred populations contain greater numbers of heterozygotes than Hardy-Weinberg equilibrium predicts.

TOPIC 6: GENETIC DRIFT

KEY POINTS

✓ *What is genetic drift?*

✓ *What is the effect of random fluctuation on populations?*

Recall from Chapter 2 that random fluctuations can cause small samples, like those in some Mendelian analyses, to deviate from expectation solely as a result of chance. This small sample size is one reason why Mendelian traits might fail a chi-square test for significance. There is a similar situation, vis-à-vis chance, in population genetics. Small populations can experience large random changes in allelic frequencies from generation to generation, solely as a result of chance. This is called **genetic drift**. Genetic drift is a random process, and it causes genotypic frequencies to deviate from Hardy-Weinberg equilibrium.

Imagine several small isolated populations that initially have the same allelic frequencies. The allelic frequencies may drift in the same direction or opposite directions at any point in time (**Figure 10.1a**). The direction of drift may also reverse abruptly, to make the populations more or less similar and to make each population more or less homozygous. All of this change happens completely at random. Furthermore, the likelihood of the number and magnitude of the changes is inversely proportional to the size of the population: The smaller the population, the more fluctuations occur and more extreme are these fluctuations (compare Figures 10.1a and 10.1b).

In extreme cases, one allele can be eliminated from the population, regardless of its phenotype. That is, the allele may be eliminated even if it is not detrimental to the organism. The likelihood that an allele will be eliminated is proportional to the size of the population and to its frequency in the population, so that the less common an allele and the smaller the population, the more likely that allele will be lost, completely by accident. The likelihood also increases the longer a population remains small. This concept is extremely important for restrictively small breeding groups, such as **population bottlenecks** (very small populations such as may occur following a natural disaster) and crops and livestock being artificially selected by breeders. If there were only two alleles of a gene before one is eliminated through genetic drift, the allele that remains is said to be **fixed** in the population ($p = 1$). After this time there can be no more fluctuation in allelic frequencies, until another allele is reintroduced by mutation, a rare event, or migration, which occurs with unpredictable frequency.

Genetic drift within a population causes a reduction in the genetic diversity of that population, restricting the gene pool. Drift occurring in related populations can easily drive them to become more similar or more different. If two populations start out at the same small size and same allelic frequencies, say $p = 0.5$, it is equally likely that they will fix to the same allele as to opposite alleles. Finally, genetic drift puts the survival of species at risk in the event of changing environmental conditions by reducing the genetic diversity of the population.

Topic Test 6: Genetic Drift

True/False

1. *Fixed* means the population has $p = 1$.
2. Genetic drift is change in allelic frequencies due to chance.

Multiple Choice

3. Which one of the following is not true? Drift occurs
   a. in small populations.
   b. in large populations.
   c. as long as one allele is not fixed.
   d. because one allele is deleterious.

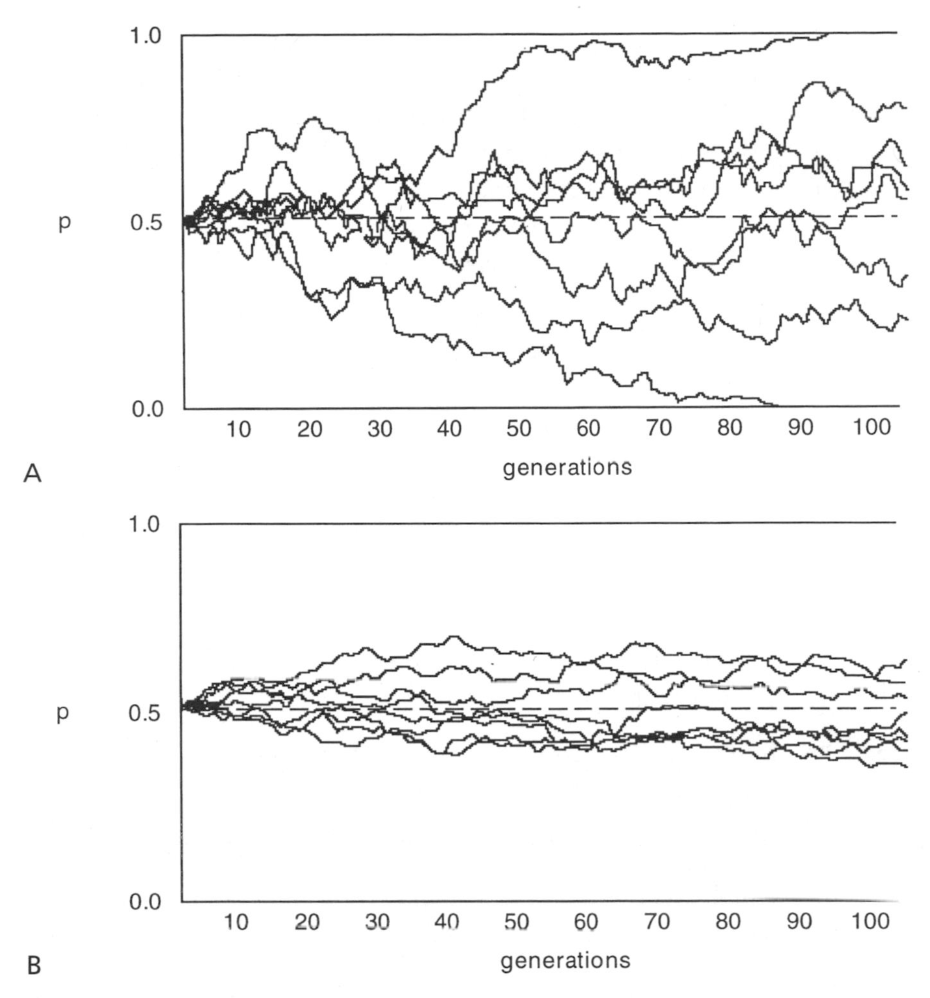

Figure 10.1 The genetic drift occurring in initially identical populations: (A) eight populations with 100 individuals each and (B) eight populations with 1,000 individuals each. Allelic frequency (p) is shown on the vertical axis and the number of generations is shown on the horizontal axis. Initially, for each population, $p = 0.5$. The horizontal line passing through the graph at $p = 0.5$ serves as a reference point for comparison of the populations—it represents what happens in the absence of genetic drift (i.e., no change in allelic frequency). To make the comparison clearer, the populations are randomly mating and are modeled without mutation, migration, and selection.

Many differences exist between the two groups of populations solely because of the 10-fold difference in size. Even the first few generations of the smaller populations (a) undergo more rapid change in allelic frequency than the larger populations (b). When populations drift away from the initial allelic frequency, the drift tends to be of greater magnitude than for larger populations. Each of the smaller populations undergoes rapid and sometimes extreme changes throughout the time period (100 generations), far more than the larger populations in the same time. The larger populations are capable of drifting this far away from initial conditions, but it takes them longer and fewer will accomplish it. A few of the smaller populations drift all the way to fixation, either of *A* ($p = 1$) or of *a* ($p = 0$). Note the direction is random. Smaller populations are more likely to drift close to fixation and then back to the initial conditions than are large populations. The larger populations remain more similar and undergo fewer and less drastic generation-to-generation fluctuations. All of these differences derive solely from the 10-fold difference in population size, which serves as a cushion against random change.

4. Which of the following populations do you expect is most likely to show genetic drift in the time specified?
   a. $p = 0.9$, 10,000 individuals, 10 generations
   b. $p = 0.9$, 1,000 individuals, 5 generations
   c. $p = 0.9$, 1,000 individuals, 2 generations
   d. $p = 0.5$, 10,000 individuals, 10 generations
   e. $p = 0.5$, 100 individuals, 5 generations

Short Answer

5. What conditions are most likely to fix one allele?
6. How does genetic drift relate to the Hardy-Weinberg law?

Topic Test 6: Answers

1. **False.** Fixed means either $p = 1$ or $p = 0$ (i.e., $q = 1$). A fixed population is described as "fixed at *A*" or "fixed at *a*."
2. **True.** The changing allelic frequencies are specifically not the result of artificial or natural selection (see Topic 7).
3. **d.** Drift occurs regardless of the phenotype of an organism bearing a particular allele.
4. **e.** The initial allelic frequency is irrelevant to the question. Figure 10.1 demonstrates the greater stability of 1,000-member populations over 100-member populations.
5. Smaller population, in which one allele is rare are most likely to fix for the other allele.
6. Genetic drift causes genotypic frequencies to randomly fluctuate, which can make them differ from the expected frequencies at Hardy-Weinberg equilibrium. The population may in fact be in Hardy-Weinberg equilibrium but the momentary fluctuation hides this fact.

TOPIC 7: NATURAL SELECTION

KEY POINTS

✓ *What is natural selection?*

✓ *How is it quantified?*

✓ *What constitutes strong or weak selection?*

✓ *How does natural selection disrupt Hardy-Weinberg equilibrium proportions?*

Natural selection is the main force driving changes in allelic frequencies and making populations more suited to their environments, a process termed **adaptation**. Natural selection occurs when one phenotype has a reproductive or survival advantage over other phenotype(s). In contrast to natural selection, migration, mutation, and genetic drift affect allelic frequencies under specific conditions and to variable extents. Nonrandom mating does not necessarily change allelic frequencies. Only natural selection will act equally on populations regardless of size to mold the genetic variability created by the other processes by changing allelic frequencies to produce adaptation in the population.

Geneticists measure the survival and reproductive success of a phenotype as **fitness**, symbolized W. Fitness is exactly reproductive success (and therefore survival to that age), nothing more, and it measures the average reproductive success for all individuals of a particular phenotype. The phenotype that produces the most offspring per individual is generally assigned a fitness of 1, and the other phenotypes are ranked relative to this one. The selection coefficient, s, expresses the degree of selection against the phenotype and is related to fitness: $s = 1 - W$. Different phenotypes can have the same or different fitness values, depending on the environmental conditions. The phenotypes with the most reproductive success have maximal fitness ($W = 1$) and minimal selection ($s = 0$). Phenotypes that completely fail to reproduce as a result of sterility or lethality have minimal fitness ($W = 0$) and maximal selection ($s = 1$).

Phenotypes can also lie anywhere along this scale. Stronger selection (higher s, lower W) produces faster change in populations. For example, a population initially balanced for two alleles of a gene, $f(A) = p = 0.5$ and $f(a) = q = 0.5$, in which the fitness of *aa* individuals suddenly plummets to 0 (it is strongly selected against), will have a q of 0.25 after only two generations. After six generations, q is already 0.125 and after many generations, the allele will essentially only be present in heterozygotes, where it can be retained through innumerable generations because the heterozygote is not selected. Usually, however, a rare allele of this sort is eliminated via inbreeding or genetic drift.

Strong selection is typified by dominant lethal alleles, which seemingly would be removed quickly from the population. The Huntington disease allele (see Chapter 3) persists because its lethality generally occurs after childbearing, and reproduction is the only consideration in fitness measurements. For other dominant lethal alleles that continually appear in the population, mutation may be a continual source of new alleles, gene interactions may cause the dominant lethal allele not to be expressed in all cases, or the allele may be incompletely penetrant. Strong selection against recessive homozygotes removes homozygotes from the population but the recessive allele remains "hidden" (i.e., undetected and unselected) in heterozygotes. Its frequency determines how rapidly it is lost: Common recessive alleles are reduced in frequency rapidly at first, more slowly later. When they finally become rare, their reduction takes longer and longer, as the matings that produce the recessive phenotype become increasingly rare.

An allele may persist for other selective reasons as well. We have already discussed the sickle-cell anemia allele (see Chapter 3), which causes a debilitating and lethal disease in homozygotes. Despite the selection against homozygotes, this allele is quite common in some areas of Africa and Asia because heterozygotes have a selective advantage over homozygous wild-type individuals. Heterozygotes are resistant to the mosquito-borne disease malaria while homozygous wild-type individuals can be seriously sickened or killed by this disease. Although homozygotes for the sickle-cell allele are relatively less fit, so are homozygous wild types in geographic areas where malaria is endemic.

Weak selection is also capable of driving alleles to extinction, especially if the allele is dominant or when the population is small (and therefore sensitive to genetic drift). Extinction is less likely or slower for recessive alleles. As the frequency of the selected-against recessive allele decreases in the population, the change in its frequency between sequential generations decreases, making it increasingly harder to eliminate the allele from the population. This is true because rare alleles exist primarily in heterozygotes, where they remain hidden and unselected.

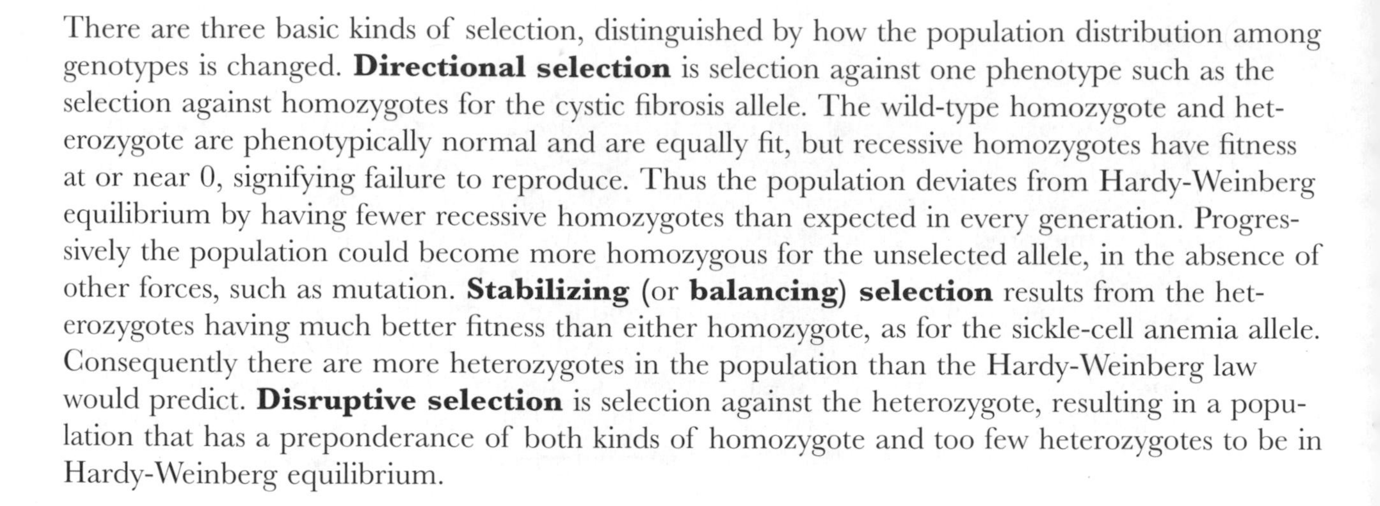

There are three basic kinds of selection, distinguished by how the population distribution among genotypes is changed. **Directional selection** is selection against one phenotype such as the selection against homozygotes for the cystic fibrosis allele. The wild-type homozygote and heterozygote are phenotypically normal and are equally fit, but recessive homozygotes have fitness at or near 0, signifying failure to reproduce. Thus the population deviates from Hardy-Weinberg equilibrium by having fewer recessive homozygotes than expected in every generation. Progressively the population could become more homozygous for the unselected allele, in the absence of other forces, such as mutation. **Stabilizing** (or **balancing**) **selection** results from the heterozygotes having much better fitness than either homozygote, as for the sickle-cell anemia allele. Consequently there are more heterozygotes in the population than the Hardy-Weinberg law would predict. **Disruptive selection** is selection against the heterozygote, resulting in a population that has a preponderance of both kinds of homozygote and too few heterozygotes to be in Hardy-Weinberg equilibrium.

Topic Test 7: Natural Selection

True/False

1. Natural selection makes populations better adapted to their environments.
2. Fitness is a constant value for each genotype.
3. Strong selection against a recessive phenotype results in rapid loss of the recessive allele.

Multiple Choice

4. A fictional fish population is composed of three sizes of fish, according to genotype, where *SS* fishes are big, *Ss* are medium sized, and *ss* are small. What sort of selection results in a population containing only large and small fishes?
   a. Directional selection
   b. Stabilizing selection
   c. Disruptive selection
   d. Weak selection
5. Which one of the following is incorrect? Natural selection
   a. changes allelic frequencies.
   b. changes genotypic frequencies.
   c. exerts an effect proportional to population size.
   d. is measured as relative fitness.
   e. is defined in terms of reproductive potential.

Short Answer

6. What is more likely to be the ultimate cause of elimination of a recessive allele whose fitness is less than 1: natural selection or genetic drift?

Topic Test 7: Answers

1. **True.** Natural selection accomplishes this feat by preventing or restricting reproduction of less-fit phenotype(s).

2. **False.** Fitness is relative and relates to phenotypes. The most reproductively successful phenotype is defined as the most fit, $W = 1$. All other phenotypes are judged relative to the most fit one. Phenotypes that do not reproduce have 0 fitness.

3. **False.** The allelic frequency initially decreases rapidly, if the allele is common, for example, $q = 0.5$. As it becomes more rare, it is lost more slowly, until it essentially undergoes no further loss (because the change is so slow) unless genetic drift eliminates it.

4. **c.** Disruptive selection reduces the number of heterozygotes in the population.

5. **c.** Unlike mutation, migration, and genetic drift, the effect of natural selection is independent of population size.

6. Genetic drift is more likely to eliminate a recessive allele than is natural selection. Natural selection can only remove recessive homozygotes from the population, leaving the recessive allele present in heterozygotes. Once the allele becomes rare, it is more likely to be lost by accident (i.e., genetic drift) because it will be mostly present in heterozygotes, who are not selected against.

IN THE CLINIC

Concepts from population genetics have been used throughout this study guide. For example, the risk of certain genetic diseases in any random mating, such as cystic fibrosis (CF), comes from knowing the allelic frequencies for those disease alleles or genotypic frequencies for heterozygotes. The likelihood of any white person being a carrier of a mutant CF allele is about 1/25. In fact, the risk estimation described in Chapter 2 In the Clinic can be made more accurate by including allele frequencies in the general population for individuals who do not have family histories of the disease. CF is quite common in white populations, relative to other populations and to other genetic diseases in this population, so the marriage of a suspected carrier to someone in the general population has a finite risk of producing an affected child. For example, consider the fictitious couple Joe and Anna. Joe's brother has CF; therefore, Joe's risk of being a carrier is 2/3 (see Chapter 2 for this calculation). Joe's wife Anna has no family history of this disease, but has a 1/25 chance of being heterozygous because she is white. If both Joe and Anna are heterozygotes, each of their children has a 1/4 chance of being affected with CF. Multiplying all of the individual risks together (see Chapter 2), we see that Joe and Anna's risk of having a child with CF is $2/3 \times 1/25 \times 1/4 = 0.0067$, or about 1/150.

Like so many alleles, the allelic frequency of the CF allele varies among ethnicities. A more accurate estimation of the likelihood of producing a child affected by CF can be known if ethnic background is taken into account. CF is most common among white populations, where the incidence among newborns is about 1/1,800. The disease is almost 10-fold less common among African-Americans: About 1/17,000 newborns is affected, reflecting the lower CF allelic frequency in this population. Asians have an even lower CF allelic frequency, having only 1/90,000 affected offspring. It should be noted that like many disease genes, the CF gene does not exhibit Hardy-Weinberg equilibrium as a result of natural selection: The recessive homozygote has lower fitness than the wild type.

Another point should be clear: No single ethnicity has more genetic disease than others. Rather, through traditions of volunteerism or as a result of well-documented pedigrees, some ethnic populations have been better studied and so their particular risks are better known at this point in time.

DEMONSTRATION PROBLEM

Question, part 1: In a particular wild dog population, two alleles of a single gene determine whether a tail is carried upright (genotype *DD*), parallel to the ground (*Dd*), or drooping (*dd*). A careful survey reveals the population to contain 408 upright tails, 900 level tails, and 480 level tails. Is this population in Hardy-Weinberg equilibrium?

Answer to part 1: The first step is to determine allelic frequencies by counting the alleles in the population. $p = f(D) = [(408 \times 2) + 900]/[2 \times (408 + 900 + 480)] = 1{,}716/3{,}576 = 0.48$ and $q = f(d) = [(480 \times 2) + 900]/[2 \times (408 + 900 + 480)] = 1{,}860/3{,}576 = 0.52$. (By calculating p and q separately, and then checking to see if their sum equals 1, you can check for arithmetical errors.) At Hardy-Weinberg (H-W) equilibrium, we expect the following:

genotype	H-W frequency	expected H-W number	observed
DD	$p^2 = 0.48^2 = 0.23$	$0.23 \times 1{,}788 = 411$	408
Dd	$2pq = 2 \times 0.48 \times 0.52 = 0.50$	$0.50 \times 1{,}788 = 894$	900
dd	$q^2 = 0.52^2 = 0.27$	$0.27 \times 1{,}788 = 483$	480

This population appears to be in Hardy-Weinberg equilibrium. This conclusion can be verified with a chi-square test, if you are not sure.

Question, part 2: A second population of these dogs has allelic frequencies $p = 0.18$ and $q = 0.82$. A group of 50 dogs from this second population joins the first population. What is the new allele frequency in the first population after the migration?

Answer to part 2: $\Delta p = m(p_m - p) = [50/(50 + 1{,}788)](0.18 - 0.48) = 0.027 \times -0.30 = -0.0081$. The new population has $p_{new} = p + \Delta p = 0.48 + (-0.0081) = 0.47$ and $q_{new} = 1 - 0.47 = 0.53$.

Chapter Test

True/False

1. New alleles in populations can result from migration.
2. To test whether a population is in Hardy-Weinberg equilibrium, we calculate q as the square root of the frequency of the recessive homozygotes.
3. Genetic drift is a natural process that affects populations relative to their sizes.
4. The value p is generally the frequency of the dominant allele.
5. Unselected changes in allele frequency are most likely for large populations.
6. A population is defined as an interbreeding group of individuals of a single species.

7. Fitness describes the relative ability of each genotype to have offspring.
8. Assortative mating causes many genes to become homozygous.

Multiple Choice

9. A laboratory population of *Drosophila* have differing eye colors, depending on their genotype for the X-linked recessive white-eye allele. Among the females, there are 137 homozygous red-eyed flies, 466 heterozygous red-eyed flies, and 397 homozygous white-eyed flies. There are also 370 red-eyed males and 630 white-eyed males. What is the frequency of the white allele in this population?
   a. 0.14
   b. 0.40
   c. 0.63
   d. 0.79
   e. 0.86
10. Dominant alleles for which $W < 1$ are often not eliminated from large populations. Which one of the following reasons cannot explain this observation?
    a. Weak selection
    b. Lethality that does not interfere with reproduction
    c. Migration from another population that contains the allele
    d. Mutation to generate new alleles
    e. All of the above are possible explanations.
11. Which one of the following is not a facet of the Hardy-Weinberg law?
    a. Rare alleles exist primarily in homozygotes.
    b. The values p^2, $2pq$, and q^2 are the observed genotypic frequencies.
    c. Mating is random.
    d. Allelic frequencies are constant.
    e. Natural selection of one phenotype is not occurring.
12. In cattle, the red, roan, and white coat colors are specified by two codominant alleles of a single gene. Roan is the heterozygous state. A particular herd is composed of 18 red, 51 roan, and 29 white animals. What is the predicted Hardy-Weinberg equilibrium frequency of the white genotype for this herd?
    a. 0.19
    b. 0.31
    c. 0.44
    d. 0.54
    e. 0.56
13. Which of the following is correct? Outbreeding is
    a. selfing.
    b. crossing to get rid of an undesirable trait.
    c. more crosses between unrelated individuals than expected from chance.
    d. matings between relatives that result in undesirable traits in offspring.
    e. another word for crossing to the wild type.
14. Mutant strains of *Caenorhabditis elegans* that have difficulty laying eggs are said to be *egl*. A particular population of 1,000 wild nematodes contains 73 *egl* individuals. What is the

predicted Hardy-Weinberg equilibrium frequency of the heterozygous genotype in this population?
a. 0.073
b. 0.20
c. 0.27
d. 0.39
e. 0.53

15. What is the most likely reason why two identical but separate populations in identical environments might have very different allelic frequencies at any point in time?
a. Mutation
b. Migration between the two populations
c. Genetic drift
d. Stabilizing selection
e. None of the above

Short Answer

16. Two incompletely dominant alleles of a single gene determine whether summer squash is long, oval, or round (oval is the heterozygous state). In one population, there are 350 plants that produce long fruit, 250 that produce oval fruit, and 400 that produce round fruit. Is this population in Hardy-Weinberg equilibrium?

17. In what ways are populations' genetic structures changed by inbreeding?

18. Yeast that are homozygous for the recessive *cyh1* allele are resistant to the drug cycloheximide. In a particular population of diploid yeast, mutations to produce *cyh1* alleles arise at a frequency around 5×10^{-4}. What will be the new frequency of the *cyh1* allele after one generation if the population initially is $q = 0.003$? (Assume the frequency of *cyh1* alleles reverting to wild type is 0.)

Essay

19. A breeder prefers white cattle, so he prevents red and roan cattle from breeding. Consequently, this herd is not in Hardy-Weinberg equilibrium. There are 9 red, 15 roan, and 63 white animals. Ownership of the herd passes to someone who mixes the cattle in a single field, allowing them to breed at will. How long will it take for a single generation to exhibit Hardy-Weinberg equilibrium? Cite allelic and genotypic frequencies in that population in your proof.

20. Describe whether and how allelic frequencies and/or genotypic frequencies are affected by nonrandom mating, mutation, migration, genetic drift, and natural selection.

Chapter Test Answers

1. **T** 2. **F** 3. **T** 4. **T** 5. **F** 6. **T** 7. **T** 8. **F** 9. **c** 10. **e** 11. **a** 12. **b** 13. **c** 14. **d** 15. **c**

16. No. Hardy-Weinberg equilibrium predicts 226 long, 499 oval, and 275 round fruit, for these allele frequencies ($p = 0.475$ and $q = 0.525$).

17. Inbreeding increases the homozygosity of many or all loci, simultaneously decreasing the heterozygous frequency of those loci. However, the allelic frequencies are not changed.

18. $q = 0.0035$.

19. Allelic frequencies in the initial population are 0.19 for the red allele and 0.81 for the white allele. Genotypic frequencies for the next population given random mating in the initial population can be demonstrated with a population Punnett square.

		red	$p = 0.19$	*white*	$q = 0.81$
red	$p = 0.19$	red	0.036	roan	0.154
white	$q = 0.81$	roan	0.154	white	0.656

These genotypic frequencies are consistent with Hardy-Weinberg equilibrium. Thus it takes one generation of random mating (in the initial population) to produce a generation that is in Hardy-Weinberg equilibrium.

20. If any of these forces affects allelic frequencies, the genotypic frequencies definitely are affected. Nonrandom mating does not affect allelic frequencies but does affect genotypic frequencies. Inbreeding increases the homozygote frequencies and outbreeding increases the heterozygote frequencies. Mutation affects allelic frequencies to a small extent. Migration has a variable effect on allelic frequencies, depending on the size of the migrating population relative to the new population and to the difference between the two populations' allelic frequencies. Genetic drift affects both allelic and genotypic frequencies. The affects are random. These affects are greatest on the smallest populations. Natural selection works on phenotypes and affects both genotypic and allelic frequencies. Stabilizing selection increases the frequency of heterozygotes. Disruptive selection increases the frequencies of both homozygotes. Directional selection decreases the frequency of one homozygote.

Check Your Performance:

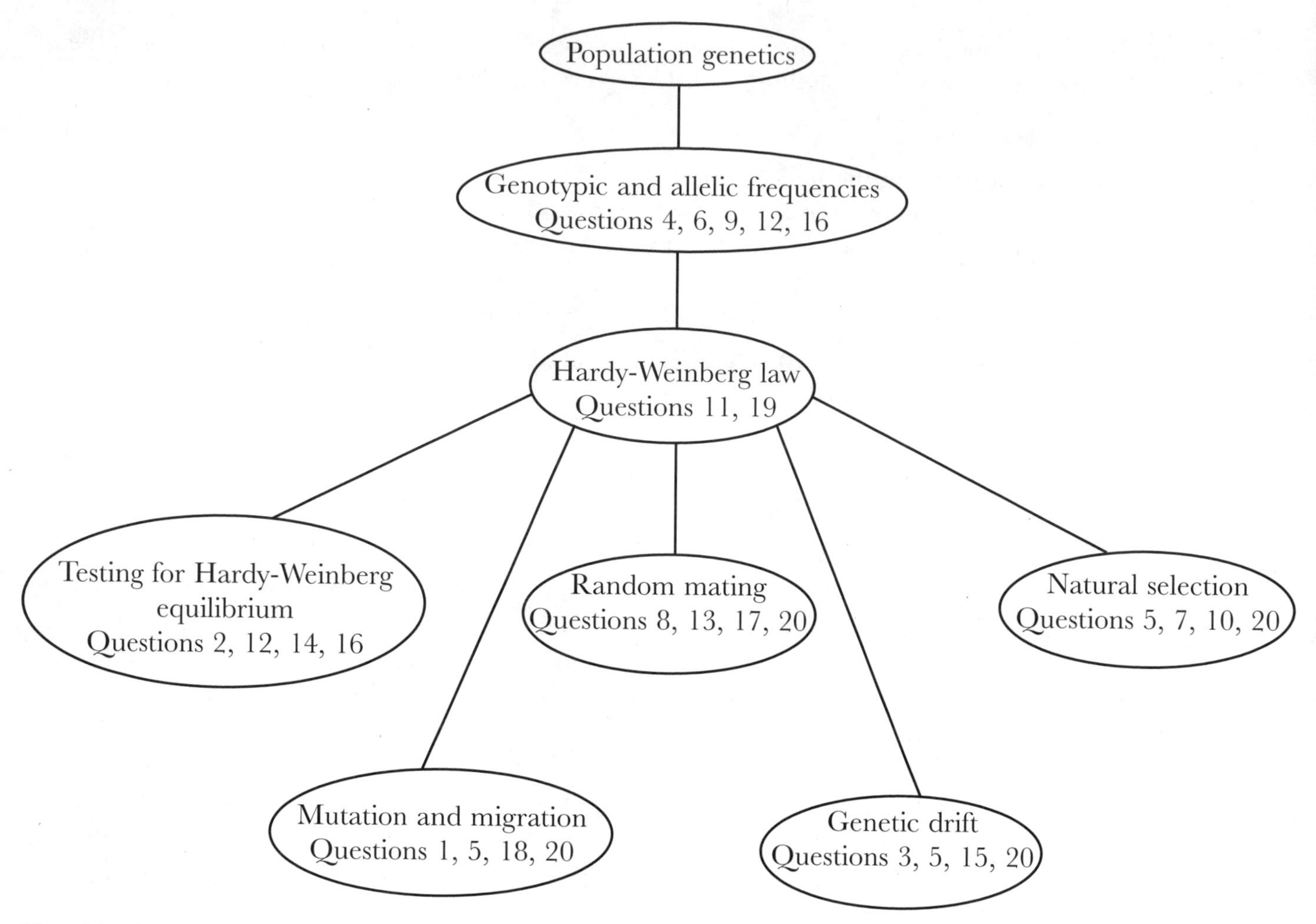

Use this chart to identify weak areas, based on the questions you answered incorrectly in the Chapter Test.

Final Exam

True/False

1. A centromere can be found in both genomic and cDNA libraries.
2. Telomeres maintain the structural integrity of the ends of chromosomes.
3. If one albino mouse is crossed to another albino mouse and all of their progeny are albino, then the parents' mutations are probably in the same gene.
4. PCR products accumulate linearly.
5. The sum rule can be used instead of the product rule to calculate probabilities of each gamete genotype from a trihybrid.
6. Populations close to fixation are more likely to fix than not to fix.
7. Monosomy is the condition of having only one homolog of a particular chromosome.
8. The offspring of a mating between a true-breeding purple corn plant and a true-breeding yellow corn plant are informative for deciding which trait is dominant.
9. Mutants for biosynthetic pathways can grow when provided with substances that occur before the block.
10. For any set of allelic frequencies, only one set of genotypic frequencies corresponds to Hardy-Weinberg equilibrium.
11. One disadvantage to fluorescence-based sequencing is that the reading and recording of the sequence are not automated.
12. Dominant alleles, like the one causing brachydactyly (short fingers), increase in frequency over time because of their dominance.

Multiple Choice

13. What are the allelic frequencies in a population consisting of 312 *AA*, 361 *Aa*, and 242 *aa* individuals?
    a. $p = 0.312$ and $q = 0.242$.
    b. $p = 0.538$ and $q = 0.462$.
    c. $p = 0.341$ and $q = 0.264$.
    d. $p = 0.462$ and $q = 0.538$.
    e. $p = 0.341$ and $q = 0.659$.
14. Which form of gene regulation is *not* present in eukaryotes?
    a. Operon gene organization
    b. Protein degradation
    c. DNA methylation
    d. Translational control
    e. mRNA processing
15. PCR is composed of a series of repeated steps, including primer annealing, primer extension, and DNA denaturation. What is the correct order for these steps?

a. Primer annealing, primer extension, DNA denaturation
b. Primer annealing, DNA denaturation, primer extension
c. Primer extension, primer annealing, DNA denaturation
d. DNA denaturation, primer annealing, primer extension
e. DNA denaturation, primer extension, primer annealing

16. If the following pathways were responsible for the appearance of elephants, what phenotype of elephant results from loss of function mutations in the *B* gene?

$$\text{smooth} \xrightarrow{A} \text{wrinkled} \xrightarrow{B} \text{crinkled} \xrightarrow{C} \text{furrowed}$$

a. Lethal
b. Smooth
c. Wrinkled
d. Crinkled
e. Furrowed

17. Which of the following is not a feature of human genetic studies?
a. Few Mendelian traits
b. Few progeny per mating
c. No planned matings
d. Unclear modes of inheritance for individual families
e. Data recorded as pedigrees

18. A diploid is formed by mating together two haploid yeast (*Saccharomyces cerevisiae*) strains: one strain is *his3 ura3* and the other is *ade2 ura3*. What medium will selectively grow the diploid (not allow the haploids to grow)?
a. Minimal medium
b. Minimal medium supplemented with adenine
c. Minimal medium supplemented with histidine
d. Minimal medium supplemented with uracil
e. Minimal medium supplemented with adenine and histidine

19. Which of the following is not a requirement in a vector?
a. A limit to the size of foreign DNA accommodated
b. Sites at which foreign DNA can be inserted
c. A means of replicating
d. A means of selecting the vector or the cells that contain it to the exclusion of others

20. The expected frequency of double crossover is
a. the difference between the frequency of both single crossovers.
b. the sum of the two single crossover frequencies.
c. the product of the two single crossover frequencies.
d. the ratio of one single crossover to the other single crossover.
e. independent of the frequencies of the single crossovers.

21. The inheritance of an X-linked dominant trait is displayed in the mouse pedigree below. Affected individuals are indicated by filled symbols. Beneath each individual is shown that animal's alleles for an RFLP linked to the trait. Which animals are crossed over between the RFLP and trait?

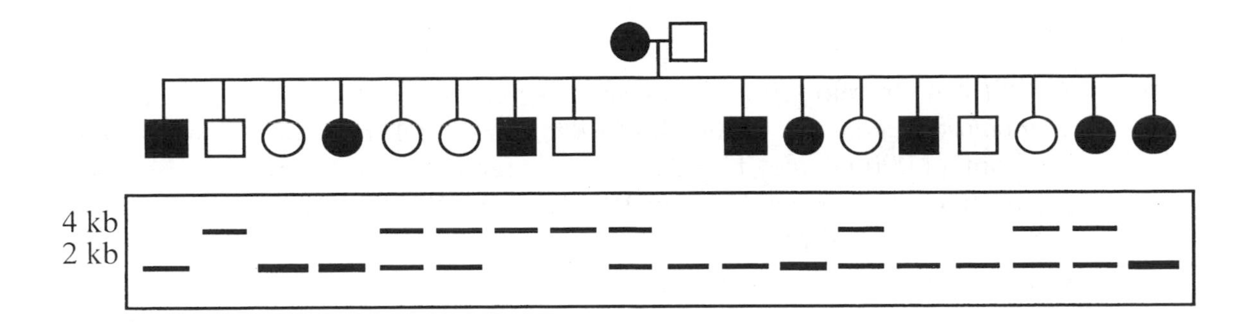

a. All of the males
b. II-3, II-4, II-10, II-16
c. II-3, II 7, II-13, II-15
d. II-1, II-4, II-10
e. I-2, II-3, II-5, II-6, II-9, II-11, II-12

22. In one population of butterflies, the frequency of a recessive allele causing stunted wings is 0.05. A second population, distinct from the first, is composed of many stunted butterflies because its allelic frequency for the same recessive allele is 0.92. A group of 50 butterflies from the second population get lost and eventually join the first population, increasing its number to 550 individuals. What is the new frequency of the stunted-wings allele in this population?
a. 0.03
b. 0.08
c. 0.09
d. 0.12
e. 0.13

Short Answer

23. Bacteria can synthesize the amino acid valine from a common metabolite, pyruvate. Mutants in one gene (*A*) that cannot make valine on their own can make valine if given 2,3-dihydroxyisovalerate or 2-keto-isovalerate. Mutants in a second gene (*B*) can make valine if given 2-acetolactate or 2,3-dihydroxyisovalerate or 2-keto-isovalerate. Mutants in a third gene (*C*) can make valine if given 2-keto-isovalerate. What is the pathway for valine synthesis? Show both the intermediates and the genes.

24. A couple were surprised to have a child affected by phenylketonuria (PKU), which is caused by a rare, recessive allele. The man's sister wants to know her risk of being heterozygous. What is the probability for her being heterozygous?

25. Describe the effects of small versus large migrating populations and similar versus dissimilar allelic frequencies on recipient populations.

26. A cloned piece of rabbit DNA gives fragments of 2 kb, 3 kb, and 9 kb when digested with the restriction enzyme *Pst*I. The restriction enzyme *Xho*I gives fragments of 6 kb and 8 kb from the same DNA clone. A double digestion of this DNA with both *Pst*I and *Xho*I produces fragments of 1 kb, 2 kb, 3 kb, and 8 kb. What restriction map is consistent with these data?

27. Describe Mendel's principles.

Essay

28. Ellis-van Creveld syndrome is one of several normally rare diseases that are found at increased frequency in the Old Order Amish of Pennsylvania, an isolated population of about 14,000 people. This disease is characterized by shortened forearms and lower legs and frequently, infant lethality. Which Hardy-Weinberg law provision(s) do you think are violated in this example?

29. What is a base analog? Why are base analogs important to geneticists?

30. Predict the complementation pattern for a collection of mutations in a known pathway. (By thinking "backward," you may gain greater understanding of complementation.) The pathway is as follows:

$$\text{aspartate} \xrightarrow{A} \beta\text{-aspartyl-phosphate} \xrightarrow{B} \text{aspartate-semialdehyde} \xrightarrow{C}$$
$$\text{homoserine} \xrightarrow{D} o\text{-phospho-homoserine} \xrightarrow{E} \text{threonine}$$

Mutations 2 and 10 map to gene *A*. Mutations 1, 4, and 8 map to gene *B*. Mutations 5 and 7 map to gene *C*. Mutation 6 maps to gene *D*. Mutations 3 and 9 map to gene *E*. Describe the complementation pattern for these mutations in words or in a chart.

Answers

1. **F** 2. **T** 3. **T** 4. **F** 5. **F** 6. **F** 7. **T** 8. **T** 9. **F** 10. **T** 11. **F** 12. **F**

13. **b** 14. **a** 15. **d** 16. **c** 17. **a** 18. **d** 19. **a** 20. **c** 21. **c** 22. **e**

23. $\text{pyruvate} \xrightarrow{B} \text{2-acetolactate} \xrightarrow{A} \text{2,3-dihydroxyisovalerate} \xrightarrow{C} \text{2-keto-isovalerate} \longrightarrow \text{valine}$

24. Either the man's parents were both heterozygous or one was heterozygous and one was homozygous wild type. (We cannot be sure which, since we do not know of any affected children.) If they were both heterozygous, the sister's probability of being heterozygous is 2/3; otherwise, her probability is 1/2.

25. Large migrating populations can have a large effect on allelic frequencies in recipient populations, especially if their allelic frequencies are dissimilar. Small migrating populations have much less effect. The more dissimilar the allelic frequencies, the greater the effect.

26.
P P X
2 3 1 8 kb
or
P P X
3 2 1 8 kb

In the figure, P represents a *Pst*I recognition sequence; X represents an *Xho*I recognition sequence. The numbers represent sizes of fragments in kilobases. Two maps are equally possible because of ambiguity in the data. *Pst*I digestion generates the 2-kb and 3-kb fragments illustrated, and a 9-kb fragment composed of the 8-kb and 1-kb fragments that are connected by a *Xho* site. *Xho*I digestion generates an 8-kb fragment, and a 6-kb fragment that comprises the 2-, 3-, and 1-kb fragments seen in the double digestion.

27. The principle of segregation states that alleles will separate from each other during the production of gametes so that they are equally transmitted to the progeny. The principle

of independent assortment states that when two or more genes segregate in a cross, the alleles of all the genes will be randomly distributed among the gametes.

28. Genetic drift in this small population may have made this and other diseases more frequent than in the rest of the world. Natural selection is acting on this population, in that Ellis-van Creveld syndrome has lower fitness than wild type (i.e., high infant lethality rates).

29. Base analogs are chemicals that are similar to regular nucleotides and can be erroneously incorporated into DNA during replication. They are important to geneticists because they mispair sufficiently often to cause base-substitution mutations.

30. Mutation 1 complements 2, 3, 5, 6, 7, 9, and 10 and fails to complement 4 and 8. Mutation 2 complements 1, 3, 4, 5, 6, 7, 8, and 9 and fails to complement 10. Mutation 3 complements 1, 2, 4, 5, 6, 7, 8, and 10 and fails to complement 9. Mutation 4 complements 2, 3, 5, 6, 7, 9, and 10 and fails to complement 1 and 8. Mutation 5 complements 1, 2, 3, 4, 6, 8, 9, and 10 and fails to complement 7. Mutation 6 complements everything but itself. Mutation 7 complements 1, 2, 3, 4, 6, 8, 9, and 10 and fails to complement 5. Mutation 8 complements 2, 3, 5, 6, 7, 9, and 10 and fails to complement 1 and 4. Mutation 9 complements 1, 2, 4, 5, 6, 7, 8, and 10 and fails to complement 3. Mutation 10 complements 1, 3, 4, 5, 6, 7, 8, and 9 and fails to complement 2.

	1	2	3	4	5	6	7	8	9	10
1	–	+	+	–	+	+	+	–	+	+
2		–	+	+	+	+	+	+	+	–
3			–	+	+	+	+	+	–	+
4				–	+	+	+	–	+	+
5					–	+	–	+	+	+
6						–	+	+	+	+
7							–	+	+	+
8								–	+	+
9									–	+
10										–